Objective Guide in Entomology

About the Authors

Dr. Devinder Sharma is a young energetic and enthusiastic scientist. He did his Master degree in Entomology and Agricultural Zoology from Banaras Hindu University. After completing PhD (Ag.) Entomology from SKUAST-Jammu. He joined Division of Entomology at SKUAST-Kashmir. Presently he is working as assistant Professor (Senior Scale) in the Division of Entomology, Sher-e-Kashmir University of Agricultural Sciences and Technology of Jammu, Jammu and Kashmir, India. He is recipient of Binani Gold Medal, BHU Gold Medal, Jawaharlal Nehru memorial Merit Certificate and Thakur Das Nicobai Kapoor Merit Scholarship during his study. He is member of various scientific societies and also associated with several journals as referee. He has over 30 national and international publications to his credit. He has co-authored one book "Honeybee mites and their management" and has written 20 book chapters. Presently he is engaged in teaching and research activities in the Division of Entomology, Sher-e-Kashmir University of Agricultural Sciences and Technology of Jammu, Jammu and Kashmir, India.

Dr. V. Kaul is an agricultural scientist with over 30 years of research and Development experience in the entomological research. His focal subject is Nematode systematics, Biocontrol and Entomopathogenic Nematodes. He has significantly contributed to the research on entomopathogenic nematodes in insect-pest management. He is presently working as Professor and Head, Division of Entomology, Sher-e-Kashmir University of Agricultural Sciences and Technology of Jammu, Jammu and Kashmir, India. He did his PhD from Kashmir University. He joined Division of Entomology, Sher-e-Kashmir University of Agricultural Sciences and Technology-Kashmir, Jammu and Kashmir, India where he served in various capacities as Assistant and then associate Professor. He is member of various scientific societies and also associated with several journals as referee. He has over 80 national and international publications to his credit.

Objective Guide in Entomology

Devinder Sharma
V. Kaul

A Paperback Division of

NEW INDIA PUBLISHING AGENCY

101, Vikas Surya Plaza, CU Block, LSC Market
Pitam Pura, New Delhi 110 034, India
Phone: + 91 (11)27 34 17 17 Fax: + 91(11) 27 34 16 16
Email: info@nipabooks.com
Web: www.nipabooks.com

Feedback at feedbacks@nipabooks.com

ISBN: 978-93-86546-10-4

Preface

The book entitled "Objective Guide in Entomology" embodies a fairly comprehensive treatment of the objective type questions on Basic and Applied Agricultural Entomology. This book covers the entire field of entomology step by step in the form of questions and answers to be specifically useful for interviews viva-voce comprehensive preliminary examinations *etc.* This book consists of different objective type questions like multiple choice on the various topics like systematics, Taxonomy, Morphology, Embryology, Physiology, Ecology, Biological control Toxicology, Host plant resistance, Innovative approaches in pest control, Integrated pest management, pesticides application equipments, pests of various crops and their management, pests of stored product and their management arthropod vectors of plant diseases, apiculture, sericulture, Lac insect and other useful insects. This book will be of great help in recollecting vast knowledge within a short time to meet the requirement of students appearing in ASRB and other competitive examinations multiple choice objective type questions have also been farmed and appended at the end of the book. The book would have useful to students preparation for various competitive examinations conducted by Indian Council of Agricultural Research-JRF, Agricultural Scientific Recruitment Board-ARS/NET, National Academy of Agricultural Research Management-SRF, Indian Agricultural Research Institute-PG/Ph.D. Entrance, Union Public Service Commission, SAUs Entrance and Allied Agricultural Examinations and also useful to teachers to keep them abreast with recent developments in entomology and guide their students to a better future.

It is our great pleasure to express our sincere thanks to New India Publishing Agency (NIPA), New Delhi who whole heartedly accepted our proposal to publish this book within a short time. Criticism and suggestions for the improvement of the book will be welcome.

Jammu, India

Devinder Sharma
V. Kaul

Contents

Syllabus for ICAR (JRF), ICAR (SRF), IARI (Ph.D) and ARS/NET Examination

Syllabus for ICAR (JRF)

Code 04 : MAJOR SUBJECT GROUP "D" - ENTOMOLOGY AND NEMATOLOGY (Subjects: D-1: Agricultural Entomology, D-2: Nematology, D-3: Apiculture, D-4: Sericulture, D-5: Plant Protection.)

UNIT-I : Importance of Agriculture in national economy; basic principles of crop production; cultivation of rice, wheat, chickpea, pigeon-pea, sugarcane, groundnut, tomato, cole crops, mango, grapes, banana, oilseeds other than groundnut, soybean and mustard. Major soils of India, role of NPK and their deficiency symptoms. Mendelian genetics; elementary knowledge of photosynthesis; respiration, and transpiration; Major cropping systems (rice-wheat cropping, crop rotations, mixed cropping); soil degradation-soil salinity and acidity and management; some aspects of post-harvest technology; varietal improvement; importance of heterosis in crop production; crop protection principles in field and storage. Major insect pests and diseases of agricultural crops like rice, cotton, pulses, oilseed crops like groundnut, soybean and mustard, vegetables like tomato, cole crops; fruit crops like mango and banana and their management principles. Transgenic crops. Important rural development programmes in India; organizational set up of agricultural research, education and extension in India; Elements of statistics.

UNIT-II : Classification of animal kingdom up to class; distinguishing characters up to orders in class Insecta; general organization of an insect external morphology with special reference to lepidopteran larvae, coleopteran adults; and honeybee; metamorphosis and moulting; different physiological systems; insect plant relationship; insect pests of agricultural and horticultural crops, and their stored/processed products, insect vectors of plant diseases- identification, biology, nature of damage, and their management tactics; and pests of household, medical and veterinary importance and their control; useful and beneficial insects like honeybee, lac insect, silkworm and pollinators; Nematode taxonomy, biology of important plant parasitic nematodes and their control; entomopathogenic nematodes, basic principles of insect and nematode pest management-cultural, biological, insecticidal, quarantine, and regulatory aspects; insecticide classification and insecticide resistance management; and insect protective transgenic crops.

ICAR SRF(PGS)

Unit 1: Systematics

History and development of Entomology, Evolution of insects, position of insects in the animal world, characteristics of phylum Arthropoda, structural features of important arthropod groups such as Trilobita, Chelicerata and Mandibulata, structural features of important classes of phylum Arthropoda viz. Arachnida, Crustacea, Chilopoda, Diplopoda and Hexapoda. Classification of insects up to order level, habits, habitats and distinguishing features of different Order and important Families.

Unit 2: Morphology

Body wall, its structure, outgrowths, endoskeleton, Body regions, segmentation, sclerites and sutures, Insect Colors. Head and head appendages, types of mouth parts, antennae, their structure and types. Thorax structure, thoracic appendages and their modification.Wings, their modification and venation,Abdomen; structure, abdominal appendages both in Pterygota and Apterygota. External genitalia, general structure and modification in important insect orders.

Unit 3: Embryology, Internal Anatomy and Physiology

Embryonic and post embryonic development, types of metamorphosis, physiology of ecdysis. General features and types of larvae and pupae. Structure, function and physiology of Digestive, Circulatory, Respiratory, Reproductive, Nervous and Excretory systems, Sense Organs; structure and types. Insect food and nutrition; minerals, carbohydrates, proteins and amino acids, lipids, vitamins and their role in growth and development, artificial diets.

Unit 4: Ecology

Concept of ecology, Environment and its components-biotic and abiotic factors and their effects on growth, development, population dynamics, distribution and dispersal. Principle of biogeography and insects biodiversity. Assessment of diversity indices. Biotic potential and environmental resistance. Ecosystems, agro-ecosystems analysis, their characteristics and functioning. Intra and inter specific relationship; competition, predator-prey and host-parasite interactions, ecological niche. Life table studies, population models. Food chain and food web. Arthropod population monitoring, pest forecasting. Diapause and causes of pest out breaks.

Unit 5: Biological Control

Importance and scope of biological control, history of biological control: Biocontrol agents-parasites, predators and insect pathogens. Important entomophagous insect Orders and Families. Ecological, biological, taxonomic, legal and economic aspects of biological control, phenomena of multiple parasitism, hyperparasitism, superparasitism and their applied importance. Principles and procedures of using exotic biocontrol agents. Utilization of natural biocontrol agents: conservation, habitat management and augmentation. Mass multiplication techniques and economics. Effective evaluation techniques, Biocontrol organizations in world and India. Successful cases of biological control of pests. Use of biotechnological tools in enhancing the potentials of Bio-Control Agents.

Unit 6: Chemical Control and Toxicology

History, scope and principles of chemical control. Insecticides and their classification. Formulations of insecticides. Susceptibility of insects to the entry of insecticides. Physical, chemical and toxicological properties of different groups of insecticides: chlorinated hydrocarbons, organophosphates, carbamates, synthetic pyrethroids, chlordimeform, chitin synthesis inhibitors, avermectins, nitroguanidines, phenylpyrrozzoles, botanicals (natural pyrethroids, rotenone, neem products, nicotine, pongamia spp. *etc*). Chloronicotinyl, pyrozole, phenyl-pyrrozzoles, oxadiazines, benzamidazole, neristoxin, rodenticides, insect hormones, Insecticide induced resurgence. Combination insecticides. Problems of pesticide hazards and environmental pollution.Safe use of pesticides, precautions and first aid treatments. Insecticides Act 1968, registration and quality control of insecticides. Evaluation of toxicity, methods of toxicity testing, determination of LD_{50}, LT_{50}, RL_{50} etc. Pesticides residues in the environment and their dynamics of movements, methods of residue. Pharmacology of insect poisons. Mode of action of different groups of insecticides; neuroactive (axonal and synaptic) poisons, respiratory poisons, chitin synthesis inhibitors. Metabolism of insecticides; activative and degradative metabolism, detoxification enzymes and their role in metabolism. Selectivity of insecticidal actions; insecticide resistance; mechanism, genetics and management of insecticide resistance.

Unit 7: Host Plant Resistance

Chemical ecology: mechano- and chemoreceptors. Host plant selection by phytophagous insects. Secondary plant substances and their defenses against phytophagous insect. Basis of resistance (Antixenosis, Antobiosis, Tolerance). Biotypes development and its remedial measures. Tritrophic interactions, induced resistance. Breeding for insect resistant plant varieties. Resistance development and evaluation techniques. Genetics of Resistance: vertical resistance, horizontal resistance, oligogenic resistance, polygenic resistance. Biotechnological approaches and development of transgenic insect resistant plants, its advantages and limitations. Case histories, insect resistance to transgenic plants and its management.

Unit 8: Innovative Approaches in Pest Control

Behavioral control: pheromones-types and uses, advantages and limitations. Hormonal control: types and function of insect hormones, insect hormone mimics, advantages and limitations. chemosterilants, antifeedants, attractants, repellents; their types, method of applications, advantages and limitations. Genetic control: concepts and methods, case histories, advantages and limitations. Potentialities of IPM.

Unit 9: Integrated Pest Management

History, concept and principles of IPM. Components of IPM: Host plant resistance, agronomic manipulations, mechanical and physical methods, chemical methods, biocontrol agents utilization, genetic and behavioral control strategy *etc*. IPM strategies for field and horticultural crops. IPM case histories, Concept of damage levels- Economic threshold levels (ETL), Economic injury levels (EIL) and their determination. System approach, Agro ecosystem and cropping system vs. IPM. Constraints and Strategies of IPM implementation. Plant quarantine laws and regulations.

Unit 10: Pesticide Application Equipments

Types of appliances: sprayers, dusters, fog generators, smoke generators, soil injecting guns, seed treating drums, flame throwers, etc. Power operated sprayers and dusters. Types of nozzles and their uses. Maintenance of appliances. Aerial application of pesticides, principles of aerial application, factors affecting the effectiveness of aerial application. Equipments for aerial applications. Advantages and disadvantages of aerial application.

Unit 11: Pests of Field Crops and their Management

Distribution, host range, biology and bionomics, nature of damage and management of arthropod pests of cereals, Oilseed, pulses and fibre crops, sugarcane and tobacco. Polyphagous pests: locusts, termites, hairy catepillars, cut worms and white grubs.

Unit 12: Pests of Horticultural Crops and their Management

Distribution, host range, biology and bionomics, nature of damage and management of arthropod pests of vegetables, fruits and plantation crops, spices, condiments and ornamentals, Vertebrate Pests.

Unit 13: Pests of Stored Products and their Management

Fundamentals of storage of grains and grain products. Storage losses, sources of infestation/infection, factors influencing losses, insect and non-insect pests, their nature of damage and control. Microflora in storage environment and their control. Storage structures, bulk storage and bag storage, their relative efficacy and demerits. Grain drying methods and aeration. Non-insect pests (rodents, birds, mites) of stored products and their control. Regulated and quarantine pests. Integrated management of storage pests.

Unit 14: Arthropod Vectors of Plant Diseases

Common arthropod vectors *viz*., aphids, leaf hoppers, plant hoppers, whiteflies, thrips, psylids, beetles, weevils, flies, bees and mites and their relationship with the plant pathogenic fungi, bacteria, viruses, mycoplasma. Mechanism of pathogen transmission : Active mechanical transmission, biological transmission. Toxicogenic insects, mites and phytotoxemia. Some important arthropod vector transmitted diseases and their epidemiology in India. Management of vector and its effect on control of diseases.

Unit 15: Honey Bees and Bee-keeping

Honey bees and their economic importance. Bee species, their behaviour, habit and habitats. Bee Keeping: bee pasturage, hives and equipments, seasonal management. Bee enemies including diseases and their control.

Unit 16: Silkworms and Sericulture

Silkworm species, their systematic position and salient features. Rearing techniques of mulberry - muga, eri and tassar silkworms. Nutritional requirements of silkworms. Sericulture: rearing house and appliancccs, silkworm breeds, principles of voltism and nivoltism, seed production and its economics. Different molecular approaches in developing silkworm breeds. Silkworm genomics- a model genetic system- transgenic silkworm- production of foreign proteins. Mulberry pests, diseases and their management. By products of sericulture and its value addition, uses in pharmaceutical industry. Enemies and diseases of silkworms and their management. Sericulture organization in India.

Unit 17: Lac Insect

Lac insect, its biology, habit and habitats. Host Trees: pruning, inoculation, lac cropping techniques and harvesting. Enemies of lac insect and their control.

Unit 18: Other Useful Insects

Pollinators, biocontrol agents of weeds, soil fertility improving agents, scavengers. Use of insects and insect products in medicines. Usefulness of insects in scientific investigations, insects as food.

Unit 19: Statistics and Computer Application

Frequency distribution, mean, mode and median. Standard, normal, bionomial and Poisson's distribution, Sampling methods and standard errors. Correlation and regression: Partial and multiple, tests of significance; t, F, chi- square, Duncan's multiple range tests. Design of experiments: Principles of Randomized block design, Completely randomized block design, Latin square design, Split-plot designs, Probit analysis. Use of software packages like SPSS, SAS, *etc*. for the above tests and designs of experiments for analysis.

IARI (Ph.D Entomology)

Position of insects in animal kingdom - their origin, phylogeny and distribution; history and basis of insect classification; distinguishing characters of insect Orders and economically important families; concept of species and speciation; rules and regulations of zoological nomenclature; morphology-external and internal; embryonic and post-embryonic development. Insect ecology-biotic potential, biotic and abiotic resistance, effect of temperature, humidity and light on insect development and population dynamics; diapause, food chain, migration and dispersal. Fundamentals of insect physiology, different systems, their structure and function, metabolism, sense organs, insect behaviour, host plant relationship. Social and other beneficial insects; pests of field crops and stored food; principles of pest control; classification, mode of action and metabolism of insecticides; insecticidal residues; resistance and resurgence; parasites, predators and pathogenic microorganisms of crop pests, biological control. Antifeedants, hormones, growth regulators, semiochemicals, host-plant resistance and genetic manipulation, insect quarantine; concept of integrated pest management; non-insect pests and their control.

ARS/NET Examination

Unit 1: Systematics

History and development of Entomology, Evolution of insects, position of insects in the animal world, characteristics of phylum Arthropoda, structural features of important arthropod groups such as Trilobita, Chelicerata and Mandibulata, structural features of important classes of phylum Arthropoda *viz.* Arachnida, Crustacea, Chilopoda, Diplopoda and Hexapoda. Classification of insects up to order level, habits, habitats and distinguishing features of different Order and important Families.

Unit 2: Morphology

Body wall, its structure, outgrowths, endoskeleton, Body regions, segmentation, sclerites and sutures. Head and head appendages, types of mouth parts, antennae, their structure and types. Thorax structure, thoracic appendages and their modification.Wings, their modification and venation, Abdomen; structure, abdominal appendages both in Pterygota and Apterygota. External genitalia, general structure and modification in important insect orders.

Unit 3: Embryology, Internal Anatomy and Physiology

Embryonic and post embryonic development, types of metamorphosis, physiology of ecdysis. General features and types of larvae and pupae. Structure, function and physiology of Digestive, Circulatory, Respiratory, Reproductive, Nervous and Excretory systems, Sense Organs; Structure and types. Insect food and nutrition; minerals, carbohydrates, proteins and amino acids, lipids, vitamins and their role in growth and development, artificial diets.

Unit 4: Ecology

Concept of ecology, Environment and its components-biotic and abiotic factors and their effects on growth, development, population dynamics, distribution and dispersal. Principle of biogeography and insects biodiversity. Biotic potential and environmental resistance. Ecosystems, agroecosystems analysis, their characteristics and functioning. Intra and inter specific relationship; competition, predator-prey and host-parasite interactions, ecological niche. Life table studies, population models. Food chain and food web. Arthropod population monitoring, pest forecasting. Diapause and causes of pest out breaks.

Unit 5: Biological Control

Importance and scope of biological control, history of biological control: Biocontrol agents-parasites, predators and insect pathogens. Important entomophagous insect Orders and Families. Ecological, biological, taxonomic, legal and economic aspects of biological control, phenomena of multiple parasitism, hyperparasitism, superparasitism and their applied importance. Principles and procedures of using exotic biocontrol agents. Utilization of natural biocontrol agents: conservation, habitat management and augmentation. Mass multiplication techniques and economics. Effective evaluation techniques, Biocontrol organizations in world and India. Successful cases of biological control of pests. Entomophilic pathogens: bacterial, fungi, viruses, rickettsiae, Protozoan and nematodes, Modes of transmission, methods of uses, symptoms of infection. Microbial insecticides and their formulation. Merits and demerits of microbial control. Role of biocontrol agents and microbial insecticides in Integrated Pest Management.

Unit 6: Chemical Control and Toxicology

History, scope and principles of chemical control. Insecticides and their classification. Formulations of insecticides. Susceptibility of insects to the entry of insecticides. Physical, chemical and toxicological properties of different groups of insecticides: chlorinated hydrocarbons, organophosphates, carbamates, synthetic pyrethroids, chlordimeform, chitin synthesis inhibitors, avermectins, nitroguanidine, phenylpyrazoles, botanicals (natural pyrethroids, rotenone, neem products, nicotine, pongamia spp. *etc*). Combination insecticides. Problems of pesticide hazards and environmental pollution. Safe use of pesticides, precautions and first aid treatments. Insecticides Act 1968, registration and quality control of insecticides. Evaluation of toxicity, methods of toxicity testing, determination of LD_{50}, LT_{50}, RL_{50} *etc*. Pesticides residues in the environment and their dynamics of movements, methods of residue. Pharmacology of insect poisons. Mode of action of different groups of insecticides; neuroactive (axonal and synaptic) poisons, respiratory poisons, chitin synthesis inhibitors. Metabolism of insecticides; activative and degradative metabolism, detoxification enzymes and their role in metabolism. Selectivity of insecticidal actions; insecticide resistance; mechanism, genetics and management of insecticide resistance.

Unit 7: Host Plant Resistance

Chemical ecology: mechano and chemo receptors. Host plant selection by phytophagous insects. Secondary plant substances and their defenses against phytophagous insect. Basis of resistance (Antixenosis, Antobiosis, Tolerance). Biotypes development and its remedial measures. Tritrophic interactions, induced resistance. Breeding for insect resistant plant varieties. Resistance development and evaluation techniques. Genetics of Resistance: vertical resistance, horizontal resistance, oligogenic resistance, polygenic resistance. Biotechnological approaches and development of transgenic insect resistant plants, its advantages and limitations. Case histories, Insect resistance to transgenic plants and its management.

Unit 8: Innovative Approaches in Pest Control

Behavioral control: pheromones-types and uses, advantages and limitations. Hormonal control: types and function of insect hormones, insect hormone mimics, advantages and limitations, chemosterilants, antifeedants, attractants, repellents; their types, method of applications, advantages and limitations. Genetic control: concepts and methods, case histories, advantages and limitations. Potentialities of IPM.

Unit 9: Integrated Pest Management

History, concept and principles of IPM. Components of IPM: Host plant resistance, agronomic manipulations, mechanical and physical methods, chemical methods, biocontrol agents utilization, genetic and behavioral control strategy *etc*. IPM strategies for field and horticultural crops. IPM case histories, Concept of damage levels- Economic threshold levels (ETL), Economic injury levels (EIL) and their determination. System approach, Agro ecosystem and cropping system vs. IPM. Constraints and Strategies of IPM implementation.

Unit 10: Pesticide Application Equipments

Types of appliances: sprayers, dusters, fog generators, smoke generators, soil injecting guns, seed treating drums, flame throwers, etc. Power operated sprayers and dusters. Types of nozzles and their uses. Maintenance of appliances. Aerial application of pesticides, principles of aerial application, factors affecting the effectiveness of aerial application. Equipments for aerial applications. Advantages and disadvantages of aerial application.

Unit 11: Pests of Field Crops and their Management

Distribution, host range, biology and bionomics, nature of damage and management of arthropod pests of cereals, Oilseed, pulses and fibre crops, sugarcane and tobacco. Polyphagous pests: locusts, termites, hairy catepillars, cut worms and white grubs.

Unit 12: Pests of Horticultural Crops and their Management

Distribution, host range, biology and bionomics, nature of damage and management of arthropod pests of vegetables, fruits and plantation crops, spices, condiments and ornamentals.

Unit 13: Pests of Stored Products and their Management

Fundamentals of storage of grains and grain products. Storage losses, sources of infestation/infection, factors influencing losses, insect and non-insect pests, their nature of damage and control. Microflora in storage environment and their control. Storage structures, bulk storage and bag storage, their relative efficacy and demerits. Grain drying methods and aeration. Non-insect pests (rodents, birds, mites) of stored products and their control. Integrated management of storage pests.

Unit 14: Arthropod Vectors of Plant Diseases

Common arthropod vectors *viz*., aphids, leaf hoppers, plant hoppers, whiteflies, thrips, psylids, beetles, weevils, flies, bees and mites and their relationship with the plant pathogenic fungi, bacteria, viruses, mycoplasma. Mechanism of pathogen transmission: Active mechanical transmission, biological transmission. Toxicogenic insects, mites and phytotoxemia. Some important arthropod vector transmitted diseases and their epidemiology in India. Management of vector and its effect on control of diseases.

Unit 15: Honey Bees and Bee-keeping

Honey bees and their economic importance. Bee species, their behaviour, habit and habitats. Bee Keeping: bee pasturage, hives and equipments, seasonal management. Bee enemies including diseases and their control.

Unit 16: Silkworms and Sericulture

Silkworm species, their systematic position and salient features. Rearing techniques of mulberry-muga-eri and tassar silkworms. Nutritional requirements of silkworms. Sericulture: rearing house and appliances, silkworm breeds, principles of voltism and nivoltism, seed production and its economics. Enemies and diseases of silkworms and their management. Sericulture organization in India.

Unit 17: Lac Insect

Lac insect, its biology, habit and habitats. Host Trees: pruning, inoculation, lac cropping techniques, and harvesting. Enemies of lac insect and their control.

Unit 18: Other Useful Insects

Pollinators, biocontrol agents of weeds, soil fertility improving agents, scavengers. Use of insects and insect products in medicines. Usefulness of insects in scientific investigations, insects as food.

Unit 19: Statistics and Computer Application

Frequency distribution, mean, mode and median. Standard, normal, bionomial and Poisson's distribution, Sampling methods and standard errors. Correlation and regression: Partial and multiple, tests of significance; t, F, chi- square, Duncan's multiple range tests. Design of experiments: Principles of Randomized block design, Completely randomized block design, Latin square design, Split-plot designs. Probit analysis. Use of software packages like SPSS, SAS, *etc.* for the above tests and designs of experiments for analysis.

Unit-1

Insect Morphology

MCQ's

1. In cockroach, collateral glands helps in
   (A) Couplation (B) Fertilization
   (C) **Formation of ootheca** (D) Formation of genital chamber

2. Dorsal surface of insect is known as
   (A) Pleuron (B) Sternum
   (C) **Tergum** (D) Plastron

3. In insects, forewings are articulated at
   (A) Prothorax (B) **Mesothorax**
   (C) Metathorax (D) 1st Abdominal segment

4. Study of various parts of mouthparts of insects reveals
   (A) Homologus organs (B) **Analogus organs**
   (C) Retrogression (D) Evolution

5. Which of the following will not under go moulting
   (A) Outer cuticle (B) **Tracheal end cell**
   (C) Tracheal trunk (D) Trachea

6. The author of "The principles of insect morphology"
   (A) A. D. Imms (B) R. F. Chapman
   (C) W. B. Wigglesworth (D) **R. E. Snodgrass**

7. Doufour's Gland, a pheromone secreting gland is found in
   (A) Male cockroach (B) Worker honeybee
   (C) **Worker ant** (D) Bark beetle

8. The visual organs present in holometabolus larvae is
   (A) Dorsal Ocelli (B) Compound eye
   (C) **Stemmata** (D) Both compound eyes and dorsal ocelli

9. In Katydids, tympanum is located at
   (A) Tibium of hind leg (B) Tibia of middle leg
   (C) **Tibia of foreleg** (D) Pleuron of 1st Abdominal segment

10. In typical insects, the number of retinula cells in an ommitidium are
   (A) Two (B) **Four**
   (C) Six (D) Eight

11. Johnston's organ is located at
   (A) The base of foreleg (B) **Second antennal segment**
   (C) First antennal segment (D) First abdominal segment

12. Proprioceptors are associated with the sense of
   (A) **Internal body changes** (B) Chemicals
   (C) Temperature (D) Touch

13. The type of larva of neuropteran is
   (A) Scarabaeform (B) **Campodeiform**
   (C) Eucepahlus (D) Apodus

14. Physogastry is a physiological phenomenon generally occurs in
   (A) Workers termite (B) **Female termite**
   (C) Male termite (D) Soldier termites

15. Tentorium is
   (A) A type of leg (B) Exoskeleton of head capsule
   (C) A type of antennae (D) **Endoskeleton of head capsule**

16. Gula is present in insect order
   (A) **Coleoptera** (B) Diptera
   (C) Lepidoptera (D) Odonata

17. The eggs present in ootheca of cockroach is
   (A) **16** (B) 10
   (C) 20 (D) 24

18. The food channels in siphoning type mouthpart is formed by
   (A) **Modified galea** (B) Labium
   (C) Maxillary palpi (D) Hypopharynx

19. The cement layer of the epicuticle is formed by the secretion of
   (A) Trichogen cells (B) Dermal glands
   (C) **Epidermal cells** (D) None of the above

20. Exarate type pupa is common in
    (A) Lepidoptera (B) Diptera
    (C) **Coleoptera** (D) Hymenoptera

21. Alula is situated at the base of
    (A) Halteres (B) Last abdominal segment
    (C) **Forewing of Diptera** (D) Forewing of strepsiptera

22. The female external genetilia of grasshopper are borne on the
    (A) **8th and 9th Abdominal segment**
    (B) 8th abdominal segment
    (C) 9th and 10th Abdominal segment
    (D) 9th abdominal segment

23. An insect leg is
    (A) Alecithal (B) Mesolecithal
    (C) Metalecithal (D) **Centrolecithal**

24. The newly hatched silkworm larva is called
    (A) Nagahara (B) **Kengo**
    (C) Denier (D) Drongo

25. The Insect of which order among these lay eggs in cocoon
    (A) Phasmida (B) Dictyoptera
    (C) Orthoptera (D) **Both A & B**

26. Insects with wings similar in structure an texture that are shed along a definite suture and have established social life
    (A) Neurptera (B) Embioptera
    (C) **Isoptera** (D) Hymenoptera

27. Insect larvae with stout sub cylindrical, C-shaped having short thoracic legs but no abdominal legs are termed as
    (A) Eruciform larva (B) Campodeiform
    (C) **Scarabaeiform** (D) Poplypodus larva

28. The wing veination nomenclature was given by
    (A) Imms, Richard and Davies (B) **Comstock and Needham**
    (C) Carlous Linnaeus (D) Martynov and Tillyard

29. The R- Cross vein of an insect wing lies between
    (A) Radial and Medius sector (B) Medius and Cubitus
    (C) Costa and Sub Costa (D) **Forks of Radius sector**

30. The book " Insect structure and function" is authored by
(A) A. D. Imms (B) **R. F. Chapman**
(C) W. B. Wigglesworth (D) R. E. Snodgrass

31. The Johnston's organ is located at
(A) Scape (B) **Pedicel**
(C) Flagellum (D) Antenna

32.The Only the right mandible is functional in
(A) Protura (B) Coleoptera
(C) **Collembola** (D) Orthoptera

33. The Absence of the right mandible is characteristic of
(A) **Thysanoptera** (B) Neuroptera
(C) Mallophaga (D) Anoplura

34. The peritrophic membrane is absent in
(A) Hemiptera (B) Hemiptera
(C) Endoparasites (D) **All the above**

35. The insect procuticle is made up of
(A) **Two layers** (B) Five layers
(C) Four layers (D) Three layers

36.The 6 segmented abdomen is found in
(A) Thysanura (B) **Collembola**
(C) Protura (D) Diplura

37.The twisted winged insects found in insect order
(A) Mallophaga (B) Mecoptera
(C) **Strepsiptera** (D) Anoplura

38.The caterpillar lack the following senses
(A) Taste (B) **Sight and hearing**
(C) Hearing and touch (D) Touch and taste

39. The constant ratio by which the linear dimensions of head capsule larvae increases as per Dyar' law is
(A) 1.6 (B) 1.2
(C) **1.4** (D) 2.5

40. In silkworm, the silk glands are modified
(A) Labial gland (B) Collembola
(C) **Salivary gland** (D) Exuvial gland

41. Malphigain tubules are absent in

(A) Schistocerca (B) Blatta

(C) Calliphora (D) **Collembola**

42. In insects, the mesentron is developed from

(A) Periderm (B) Mesoderm

(C) **Endoderm** (D) Ectoderm

43. The filter chamber is the main feature of

(A) **Digestive system** (B) Excretory system

(C) Respiratory system (D) Circulatory system

44. Haemoglobin as respiratory pigment is not present in

(A) *Chironomus* sp. (B) ***Attractomorpha* sp.**

(C) *Gasterophilus* sp. (D) *Anisops* sp.

45. In the gut of Kalotermitidae, the symbionts that help in cellulose digestion are

(A) Virus (B) Bacteria

(C) **Flagellates** (D) Fungi

46. The type of ovarioles found in cockroach is

(A) Meroistic type (B) **Panoistic type**

(C) Telotrophic type (D) Polytrophic type

47. The anti freezing compound found in insect blood is

(A) **Glycogen** (B) Trehalose

(C) Glucose (D) All of the above

48. Cryptonephridial type malphigian tubules found in

(A) Hemiptera (B) Lepidoptera

(C) Diptera (D) **Coleoptera**

49. Paedogenesis is common in

(A) Muscidae (B) **Ceccidomyidae**

(C) Aphididae (D) Hippobascidae

50. Ecdysone is secreted by

(A) Corpora Allata (B) Mandibular glands

(C) **Prothoracic gland** (D) Corpora Cardiaca

51. The phenomenon of anamorphosis occur in

(A) Thysanura (B) Collembola

(C) **Protura** (D) Diplura

52. The common inhibitory neuro transmitter in insect nervous system is
    (A) Acetylcholine (B) **GABA**
    (C) Serotonin (D) Dopamine

53. The origin of trachea is
    (A) **Ectodermal** (B) Endodermal
    (C) Mesodermal (D) None of these

54. The Polyembrony is found in
    (A) Glossina (B) Aphids
    (C) **Parasitic Hymenoptera** (D) All of these

55. The wing coupling mechanism in Mecoptera
    (A) Amplexiform coupling (B) **Jugate coupling**
    (C) Hamulate coupling (D) Frenate coupling

56. The phenomenon of anamorphosis occur in
    (A) Thysanura (B) Collembola
    (C) **Protura** (D) Diplura

57. The compound eyes are absent in order
    (A) Thysanura (B) Collembola
    (C) **Protura** (D) Diplura

58. The pulse beetle is characterize by
    (A) **Hypermetamorphosis** (B) Anamorphosis
    (C) Both of these (D) None of these

59.The halteres are characteristic of insect order
    (A) **Diptera** (B) Lepidoptera
    (C) Dermaptera (D) Hymenoptera

60. The term Hexapoda was coined by
    (A) Linnaeus (B) **Laterille**
    (C) Imms, A. D. (D) Sharov

61.The antenna are absent in class
    (A) Crustacea (B) **Arachnida**
    (C) Onychophora (D) Myriapoda

62. The Circulatory system in insects is concerned with
    (A) **Nutrient and oxygen transport**(B) Nutrient transport
    (C) Oxygen transport (D) None of the above

63. The B. P. Uvarov propound the theory of
(A) Diapause (B) **Phase in locust**
(C) Energy utilization (D) Migration

64. The Halteres are characteristic of
(A) Strepsiptera (B) Mallophaga
(C) **Diptera** (D) Hymenoptera

65. The Cryptonephridial condition help insects to
(A) Remove excess water (B) **Conserve water**
(C) Both A and B (D) None of these

66. The lateral part of thoracic segment is
(A) Sternite (B) Tergite
(C) **Pleurite** (D) Trochantin

67. The immature stages of aquatic insects is called
(A) Tumbler (B) Nymph
(C) **Naiad** (D) Grub

68. The Pharate stage is
(A) **Apolysis to ecdysis** (B) Ecdysis to next ecdysis
(C) Apolysis to next apolysis (D) Ecdysis to apolysis

69. The Blood sugar is
(A) **Trehalose** (B) Glucose
(C) Sucrose (D) Fructose

70. The saliva of *Cimex* spp.contains
(A) Formic acid (B) **Anticoagulant**
(C) Wax secretion (D) Silk protein

71. The Ptilinum is found in
(A) Hymenopterans (B) **Cyclorrhaphous Dipteran**
(C) Homoptera (D) Odonata

72. The most important adult character of insect
(A) Exocuticle (B) **Hexapod locomotion**
(C) Segmented body (D) Open circulatory system

73. The Sclerotization is controlled by
(A) Juvenile Hormone (B) Ecdysone
(C) **Bursicon** (D) Brain Hormone

74. The chitin is a polymer of

(A) **Acetyl D Glucosamine** (B) Glycogen

(C) Amino acid (D) Glucose

75.The Usually the male reproductive part are located at

(A) 7th abdominal segment (B) 8th Abdominal segment

(C) 10th abdominal segment (D) **9th Abdominal segment**

76. The Hypognathous head is found in

(A) Rice weevil (B) **Locust**

(C) Red cotton bug (D) Aphid

77. The mandibles in honeybee are used for

(A) **Nest construction** (B) foraging

(C) Cleaning (D) Feeding

78. Marcello Malphigi after whose name malphigian tubules were name belong to

(A) Germany (B) **Italy**

(C) USA (D) France

79.The honey chamber is modified form of

(A) Gizzard (B) Pharynx

(C) **Crop** (D) Midgut

80.The amount of chitin in insect cuticle is

(A) 20% (B) **40%**

(C) 25% (D) 30%

81. The respiratory organ book lungs are present in class

(A) **Crustacea** (B) Arachnida

(C) Onychophora (D) Myriapoda

82. The true fly is

(A) Whitefly (B) Dragonfly

(C) **House fly** (D) Scropionfly

83.In adult stage four pair of legs are found in

(A) Diplopoda (B) Arachnida

(C) Myriapoda (D) **Crustacea**

84. Grasping type of legs are found in

(A) Biting lice (B) **Mantids**

(C) Katydids (D) Grasshopper

85. The Communication "Dance language" in honeybees was described by

(A) R. E. Snodgrass (B) **Karl Von Frisch**

(C) T. Needham (D) V.B. Wigglesworth

86. The insects are

(A) Capable of moving the eyes (B) Incapable of moving the eyes

(C) **Both of these** (D) None of these

87. The Juvenile hormone are absent in which stage

(A) **Egg** (B) Larva

(C) Pupa (D) Adult

88. In Weismann's rings, which organ is present

(A) Nephridial system (B) **Neuron**

(C) Plasmatocytes (D) Oenocytes

89. The Weismann's ringsis formed by

(A) Corpora cardiaca (B) Corpora allata

(C) Prothoracic glands (D) **All of these**

90. The chemical essential for insect vision is

(A) Myrosine (B) Trypsin

(C) Tyrosine (D) **Rhodopsin**

91. The moulting hormone (Ecdysone) is secreted by

(A) Corpora cardiaca (B) Corpora allata

(C) **Prothoracic gland** (D) All of these

92. The Amplexiform wing coupling is present in

(A) **Papilionidae** (B) Noctuidae

(C) Aphididae (D) Pyralidae

93. The Hamulate wing coupling is found in

(A) Ephemeroptera (B) **Diptera**

(C) Mecoptera (D) Strepsiptera

94. The elytra is characteristics of

(A) Neuroptera (B) **Coleoptera**

(C) Dermaptera (D) Mallophaga

95. The mouthparts of thrips are

(A) Symmetrical (B) **Asymmetrical**

(C) Sponging type (D) Siphoning type

96. The moulting hormone (Ecdysone) is secreted by
    (A) Corpora cardiaca (B) Corpora allata
    (C) Prothoracic gland (D) All of these

97. The condition when all the ten pair of spiracles are functional
    (A) Hemipneustic (B) **Holopneustic**
    (C) Peripneustic (D) Metapneustic

98. The largest pylum in animal kingdom is
    (A) Coelentrata (B) Annelida
    (C) Arthropoda (D) Mollusca

99. Only two malphigian tubules are present in
    (A) Tettigonids (B) Acridids
    (C) **Coccids** (D) All of these

100. The sensory element of compound eye is
    (A) **Rhabdom** (B) Iris cell
    (C) Lens (D) serotonin

101. The chamber immediately below the spiracle is
    (A) Trachea (B) **Atrium**
    (C) Taenidia (D) None of these

102. Only four functional legs are present in
    (A) Noctuidae (B) Pyralidae
    (C) Bombidae (D) **Nymphalidae**

103. Paired gonophore are present in
    (A) Coleoptera (B) Trichoptera
    (C) Siphonoptera (D) **Ephemeroptera**

104. Thyridium sense organ present in
    (A) **Trichoptera** (B) Strepsiptera
    (C)` Coleoptera (D) Siphonoptera

105. Which one of the following parasitic order has laterally compressed body
    (A) Protura (B) **Strepsiptera**
    (C) Odonata (D) Mallophaga

106. The arthropod class forming connecting link between phylum Arthropda and Annelida is
    (A) Crustacea (B) Arachnida
    (C) **Onychophora** (D) Myriapoda

107. The centipedes are
   (A Herbivorous
   (B) **Carnivorous**
   (C) Omnivorous
   (D) None of these

108. The frillbugs and sowbugs belong to class
   (A) **Crustacea**
   (B) Arachnida
   (C) Onychophora
   (D) Myriapoda

109.The largest insect order is
   (A) **Coleoptera**
   (B) Hymenoptera
   (C) Lepidoptera
   (D) Diptera

110. The smallest insect is
   (A) *Megaphragma mymaripenne*
   (B) *Megaphragma caribea*
   (C) ***Dichomorpha echmepterygis***
   (D) None of these

111. The smallest insect order is
   (A) **Manophasmatodea**
   (B) Zoraptera
   (C) Mallophaga
   (D) Sterpsiptera

112. The reproductive strategies adopted by insect for dominance is/are
   (A) High fertility
   (B) High fecundity
   (C) unbalanced sex ratio
   (D) **All of above**

113. The termite queen on an average lay
   (A) **One egg/second**
   (B) One egg/minute
   (C) One egg/hour
   (D) One egg/day

114. About 88 per cent of the insects belong to
   (A) **Endopteygote**
   (B) Exopterygote
   (C) Pterygote
   (D) All of above

115. The advantage of the complete metamorphosis is
   (A) Optimized growth
   (B) compartmentalization of the life cycle
   (C) Optimized feeding
   (D) **All of above**

116. The behaviour to pretend dead to escape from enemies is known as
   (A) **Thanatosis**
   (B) Circadian rhythm
   (C) Biological rhythm
   (D) None of these

117. Communication system is well developed in
(A) Honeybee (B) Termites
(C) Ants (D) **All of above**

118. Triboliata theory of insect origin was proposed by
(A) **Handerlisch** (B) Crampton
(C) Hansen (D) Imms

119. The polyphyletic hypothesis of insect origin was proposed by
(A) Anderson (B) charlotte Manton
(C) **Both of above** (D) None of these

120. The unsclerotized area of the cuticle where actually the body wall breaks at the time of moulting is called as
(A) Sclerite (B) **Suture**
(C) Sulci (D) None of these

121. The Upper or dorsal region in case of thorax is known as
(A) Dorsum (B) Notum
(C) Tergum (D) **All of above**

122. The epidermal cells of insect cuticle are held together by
(A) Septate junctions (B) **Zonulae adherens**
(C) None of these (D) All of above

123. The chitin is present in
(A) **Epicuticle** (B) Procuticle
(C) Both of these (D) None of above

124. The Non chitinised substances of insect cuticle is
(A) Arthropodin (B) Resilin
(C) Sclerotin (D) **None of above**

125. The presence of chitin in insect cuticle can be determined by
(A) Van Wisseling test (B) Diaphenol test
(C) Chitin synthetase test (D) **All of above**

126. The rubber like protein present in wing articulation and thorax, acting like mechanical springs is
(A) Arthropodin (B) **Resilin**
(C) Sclerotin (D) None of above

127. The horny nature of exocuticle is due to
(A) Chitin (B) **Sclerotin**
(C) Phenolic substances (D) All of above

128. The Tanning in adult cuticle is caused by a peptide hormone
(A) Diuretic hormone (B) **Bursicon**
(C) Both of these (D) None of above

129. The tanning hormone Bursicon is secreted by
(A) Neuro secretory cells (NSC)
(B) Prothoracic glands (PTG)
(C) Corpora Cardiaca (CC)
(D) None of above

130. The Deep finger like invaginations of the cuticle are called as
(A) Sulci (B) **Apodemes**
(C) Tentorium (D) None of above

131. The tentorium is formed by the fusion of
(A) Two - Four apodemes (B) Three - Five Apodemes
(C) **Four - Six apodemes** (D) None of above

132. The Corporo-tentorium is present in
(A) Isoptera (B) Hymneoptera
(C) **Orthoptera** (D) Hemiptera

133. The membranous articulation of leg joints is
(A) Monocondylic (B) Dicondylic
(C) Both of these (D) None of above

134. The time interval between the two subsequent moulting is called as
(A) **Stadium** (B) Instar
(C) Moulting period (D) None of above

135. The epidermal cell forming a seta is called as
(A) **Trichogen** (B) Tormogen
(C) Both of these (D) None of above

136. The study of arrangement of seta is called as
(A) Setataxy (B) **Chaetotaxy**
(C) Seta veination (D) None of above

137. The spurs are

(A) Multicellular (B) **Unicellular**

(C) Acellular (D) None of above

138. The wax glands in honeybees are present on

(A) **4th-7th abdominal segments** (B) 5th -8th abdominal segments

(C) 2nd -5th abdominal segments (D) 3rd -6th abdominal segments

139. The Repugnatorial glands are present in

(A) Beetles (B) Saturnid moths

(C) **Plant bugs** (D) Grasshoppers

140. The androconia are present in

(A) Beetles (B) **Lepidoptera**

(C) Diptera (D) Hemiptera

141. A change in colour with the angle of viewing is known as

(A) Tyndall scattering (B) Interference

(C) Diffraction (D) **Iridescence**

142. The insect head is formed by the fusion of

(A) Four segments (B) Five segments

(C) **Six segments** (D) Seven segments

143. The prognathus head orientation is found in

(A) Butterflies (B) Grasshoppers

(C) Bugs (D) **Beetles**

144. The Lateral ocelli (Stemmata) are present in

(A) Aperygotes insect nymphs (B) Exopterygotes insect nymphs

(C) **Endopterygote insect larvae** (D) All of above

145. The Binocular vision is present in

(A) Tiger beetle (B) Tabanid flies

(C) Preying mantis (D) **All of above**

146. The broadest color sensitivity among insects is found in

(A) Butterflies (B) Dipteran flies

(C) **Honey bee** (D) All of above

147. The Johnston organ (auditory) is present

(A) Scape (B) **Pedicel**

(C) Antennary socket (D) Flagelllum

148. The Bipectinate antennae are present in

(A) **Silkworm moth** (B) Mustard Sawfly
(C) Blister beetle (D) Butterfly

149. The labial palpi is

(A) Four segmented (B) Five segmented
(C) **Three segmented** (D) None of above

150. The Ligula is present in

(A) **Beetles** (B) Grasshopper
(C) Both of these (D) None of above

151. The rasping and sucking type mouth parts are present in

(A) Beetles (B) Grasshopper
(C) Earwigs (D) **Thrips**

152. The Halteres are characteristics of order

(A) Beetles (B) Grasshopper
(C) Earwigs (D) **True flies**

153. The two segmented trochantellus is present in

(A) Dragonflies (B) Damselflies
(C) Ichneumonid wasp (D) **All the above**

154. The condition whereby all the definitive number of segments differentiate in embryo is

(A) **Epimorphic** (B) Anamorphic
(C) Hypermetamorphic (D) None of above

155. The propodeum and gaster are characteristic of order

(A) **Hymenoptera** (B) Hemiptera
(C) Diptera (D) Lepidoptera

156. The pre genital organs *viz.* glue bar, retinaculum and furca are character istic of order

(A) Diplura (B) Protura
(C) Thysanura (D) **Collembola**

157. The Physogastry is characteristic of order

(A) **Isoptera** (B) Hemiptera
(C) Phasmida (D) Isoptera

158. The mustard sawfly larvae have
(A) 5 pair of prolegs (B) 6 pair of prolegs
(C) 7 pair of prolegs (D) **8 pair of prolegs**

159. The cerci is absent in
(A) Lepidoptera (B) Thysanoptera
(C) Hemiptera (D) **All the above**

160. The functional couplatory organ on the venter of second and third abdominal segment is present in
(A) **Damselflies** (B) Mayflies
(C) Earwigs (D) All the above

161. The larva of mosquito is known as
(A) Maggot (B) **Wrigglers**
(C) Tumbler (D) All the above

162. The Decticous/Exarate/Free pupae is found in
(A) Hymenoptera (B) Neuroptera
(C) Trichoptera (D) **All the above**

163. The sub imago is characteristic of insect order
(A) Dictyoptera (B) Neuroptera
(C) Trichoptera (D) **Ephemeroptera**

164. The Juvenile hormone is secreted by
(A) Prothoracic glands (B) Neuro secretory cells (NSC)
(C) Corpora Cardiaca (D) **Corpora allata**

165. The proventriculus/gizzard is lacking in
(A) Carnivorous insects (B) **Fluid feeding insect**
(C) Omnivorous insects (D) Phytophagous insect

166. The origin of midgut is
(A) Ectodermal (B) Mesodermal
(C) **Endodermal** (D) None of these

167. The origin of foregut is
(A) **Ectodermal** (B) Mesodermal
(C) Endodermal (D) None of these

168. The filter chamber is present in
(A) **Aphids** (B) Thrips
(C) Butterflies (D) Bugs

169. The malphigian tubules are absent in
(A) Butterflies (B) Thrips
(C) **Aphids and collembola** (D) Bugs

170. The insects are mainly
(A) Ureotelic (B) **Uricotelic**
(C) Ammonotelic (D) Bugs

171. The Cryptonephridial arrangement of malphigain tubules is found in
(A) Lepidoptera (B) Hemiptera
(C) Coleoptera (D) **Orthoptera**

172. The labial kidney is present in
(A) **Collembola** (B) Thysanura
(C) Both of these (D) None of these

173. The textbook "*The Insects, Structure and Function*" is authored by
(A) **R. F. Chapman** (B) R. Snodgrass
(C) V. B. Wigglesworth (D) R. Mastuda

174. The First Indian Imperial Entomologist of India was
(A) **H.S. Pruthi** (B) S. Pradhan
(C) M.S. Mani (D) N. C. Pant

175. Ganglia within each segment are linked to one another by a short medial nerve known as
(A) **Commissures** (B) Connectives
(C) Both of these (D) None of these

176. The book "Some South Indian Insects" was written by
(A) T.V. Ramakrishna Ayyar (B) Maxwell Lefroy
(C) M.S. Mani (D) **Bainbrigge Fletcher**

177. The Numerical taxonomy includes
(A) The Phenetic approach (B) The Cladistic approach
(C) **Both of these** (D) None of these

178. The term Taxonomy was given by
(A) **De Candolle** (B) Linnaeus
(C) Lamarck (D) Haeckel

179. Which one is true fly
(A) Housefly (B) **House fly**
(C) Both of these (D) None of these

180. A taxonomic category containing a single species or a monophyletic group of species which are separated from other taxa of the same rank by a decided gap is known as

(A) Species (B) **Genus**
(C) Sub species (D) Family

181. The International Code of Zoological Nomenclature consists of

(A) 90 articles (B) 77 articles
(C) **87 articles** (D) 85 articles

182. The paleopteriod insect order is

(A) **Odonata** (B) Psocoptera
(C) Plecoptera (D) Mecoptera

183. In apterygotes the mandibular articulation is

(A) **Monocondylic** (B) Dicondylic
(C) Free (D) None of these

184. The sucking lice belongs to order

(A) **Anoplura** (B) Mallophaga
(C) Mecoptera (D) Psocoptera

185. The Snakeflies belongs to order

(A) Mecoptera (B) **Neuroptera**
(C) Strepsiptera (D) Trichoptera

186. In Australia, *Sminthurus viridis* is a serious pest of

(A) Alfalfa (B) Soybean
(C) **Leucerne** (D) Berseem

187. The adults having shortest lifespan belong to order

(A) **Ephemeroptera** (B) Trichoptera
(C) Strepsiptera (D) Mecoptera

188. The insects mostly used as indicators of water purity belongs to order

(A) **Plecoptera** (B) Grylloblattodea
(C) Strepsiptera (D) Mantophasmatodea

189. The phase theory of locust was proposed by

(A) Y. Ramachndra Rao (B) Anton Sharov
(C) **Uvarov** (D) None of these

190. The most destructive locust in India is
(A) ***Locusta migratoria*** (B) *Schistocerca gregaria*
(C) *Dociostaurus moroccanus* (D) *Patanga succincta*

191. The headquarter of Locust warning organization in India is at
(A) Jaiselmer (B) Jodhpur
(C) **Faridabad** (D) Jaipur

192. The phenomenon of autotomy is a characteristic feature of order
(A) Isoptera (B) Mecoptera
(C) **Phasmida** (D) Zoraptera

193. Paired penis is present in order
(A) Embioptera (B) Dictyptera
(C) Zoraptera (D) **Dermaptera**

194. In Embioptera, the silk is produced by the glands present on
(A) tarsi of the Middle legs (B) Tarsi of all legs
(C) Tarsi of the hind legs (D) **Tarsi of the front legs**

195. In Dictyoptera, the eggs are laid in batches contained in hard protective capsule called
(A) Ootheca (B) Sheprad's purse
(C) Egg case (D) **All the above**

196. In Termites, the soldiers having median rostrum and vestigial mandibles but have well developed frontal glands are known as
(A) **Nasutes** (B) Psuedergates
(C) False workers (D) All the above

197. Pearman's organ, a sound-producing structure is present in
(A) Trichoptera (B) **Psocoptera**
(C) Mallophaga (D) Mecoptera

198. The vector of relapsing fever transmitted by *Spirochaete recurrentis* is
(A) ***P. humanus capitis*** (B) *Haematopinus tuberculatus*
(C) *Phthirus pubis* (D) *P. humanus humanus*

199. The vector of sesamum phyllody virus is
(A) *Peregrinus maidis* (B) *Bemesia tabaci*
(C) ***Orosius albicinctus*** (D) *Cestius phycitus*

200. The aphid having winter and summer migration on the same plant is

(A) *Rhophalosiphum maidis* (B) *Aphis craccivora*

(C) ***Eriosoma lanigerum*** (D) *Aphis gossypii*

201. The vector of sesamum phyllody virus is

(A) *Peregrinus maidis* (B) *Bemesia tabaci*

(C) ***Orosius albicinctus*** (D) *Cestius phycitus*

202. The Cocoon's silk in neuropteran is produced by

(A) Salivary glands (B) **Malphigian tubules**

(C) Tarsi of fore legs (D) None of these

203. The Hangingflies belongs to order

(A) Neuroptera (B) Zoraptera

(C) **Mecoptera** (D) Plecoptera

204. The Repugnatorial glands known as Osmeteria in Papiliionidae are present on

(A) **The 1st thoracic segment** (B) The I1nd thoracic segment

(C) The 1IIrd thoracic segment (D) The 1st abdominal segment

205. In loopers, prologs are present on abdominal segments

(A) 3,4,5, 6 and 10th (B) 4, 5, 6 and 10th

(C) 5, 6 and 10th (D) **6 and 10th**

206. The skippers belong to family

(A) **Hesperiidae** (B) Noctuidae

(C) Geomatridae (D) Lymantridae

207. A bladder like Ptilinum or frontal sac is present in suborder

(A) Brachycera (B) Nematocera

(C) **Cyclorrhapha** (D) None of these

208. The vector of Kala azar is

(A) *Mansonia* (B) ***Phlebotomus papatasi***

(C) *Leishmania donovani* (D) None of these

209. The Larva characterised by the presence of a sterna spatula or Breast bone mid ventrally on the thorax belong to family

(A) Tabanidae (B) Chironomidae

(C) Simulidae (D) **Ceccidomyiidae**

210. The vector of Tularaemia or ddr fly fever of man is

(A) *Aedes aegypti*
(B) ***Phlebotomus papatasi***
(C) *Tabanus speciosus*
(D) None of these

211. The Boat shape with a float on either side is a characteristic feature of

(A) *Anopheles*
(B) ***Culex***
(C) *Aedes*
(D) *Mansonia*

212. The vector *Aphidius spp.* is a

(A) **Ectoparasitiod**
(B) Endoparasitiod
(C) Both of these
(D) None of these

213. The Long horned wood boring beetle belong to family

(A) Alticidae
(B) Bruchidae
(C) Chrysomelidae
(D) **Cerambycidae**

214. The study of individual organisms or an individual species in relation to the environment is

(A) Synecology
(B) Ecology
(C) **Autoecology**
(D) **All of these**

215. A period of suspended activity in individuals occurring irrespective of season is

(A) Hibernation
(B) Diapause
(C) Aestivation
(D) **Quiescence**

216. Insects which are active at dusk

(A) Nocturnal
(B) Diurnal
(C) **Crepuscular**
(D) None of these

217. Grubs of *Trogoderma* sp. is

(A) **Photoropic**
(B) phototaxis
(C) photonegative
(D) None of these

218. Cannabalism is found in

(A) *Helicoverpa*
(B) *Chrysoperla*
(C) **Both of these**
(D) None of these

219. Genetic theory of insect population regulation was proposed by

(A) **Chitty**
(B) Milne
(C) Thomson
(D) Soloman

220. Larval diapauses is found in
(A) **Paddy stem borer** (B) *Helicoverpa*
(C) Cabbage butterfly (D) Colorado potato beetle

221. Insect pests with high reproductive rate and low survival rate are called
(A) k strategist (B) **r strategist**
(C) s strategist (D) None of these

222. Insects are
(A) **Poikilothermic** (B) Homeothermic
(C) Heterothermic (D) Polythermic

223. Vector of "Tungro" disease in paddy is
(A) ***Nephotettix virescens*** (B) *Scirpophaga incertulas*
(C) *Nilaparvata lugens* (D) *Lipaphis erysimi*

224. Which of the following is larval parasitoid
(A) ***Apanteles/Cotesia*** (B) *Tetrasticus*
(C) *Xanthopimpla* (D) *Trichogramma*

225. The nematode responsible for "Molya" disease in wheat and barley is
(A) *Helicotylenchus* sp. (B) *Meloidogyne* sp.
(C) *Trichodorus* sp. (D) ***Heterodera* sp.**

226. The causal organism of "Ufra" disease in rice is
(A) ***Ditylenchus angustus*** (B) *Heterodera avenae*
(C) *Globodera rostochinensis* (D) *Tylenchulus semipenetrans*

227. The "Ear cockle" disease in wheat is caused by
(A) *Hirschmanniella oryzae* (B) *Tylenchulus semipenetrans*
(C) *Heterodera avenae* (D) ***Anguina tritici***

228. "Tundu" or "Tannan" disease of wheat is caused by the nematode in association of
(A) Virus (B) **Bacteria**
(C) Fungus (D) Protozoa

229. The double seed in cotton is a characteristic of
(A) *Helicoverpa armigera* (B) *Eariasfabia*
(C) ***Pectinophora gossypiella*** (D) *Sylepta derogata*

230. Clipping of leaves at tip of paddy seedling helps to manage
(A) **Rice stem borer** (B) Rice butterfly
(C) Paddy gallfly (D) Brown plant hopper

231. The scientific name includes the name of a scientist in following case

(A) ***Dysdercus koenigii*** (B) *Helicoverpa armigera*

(C) *Empoasca devastans* (D) *Plutella xylostella*

232. *Bandicota bengalensis* is a

(A) House rat (B) Field rat

(C) House mouse (D) **Mole rat**

233. The "17 Year locusts" is common name of

(A) Locust (B) **Cicada**

(C) Grasshopper (D) *Chironomous*

234. One of the molluscicide is

(A) **Metaldehyde** (B) Malathion

(C) Heptachlor (D) Endosulfan

235. The identification mark generally found on the biopesticdes.

(A) Red (B) Yellow

(C) Blue (D) **Green**

236. Rope method is used for the control of

(A) **Rice leaf folder** (B) Rice stem borer

(C) Sugarcane mealybug (D) Mango stemborer

237. The colour label of extremely toxic insecticide is

(A) **Red with skull bone** (B) Yellow

(C) Green (D) Blue

238. Pollination in mango is mediated by

(A) **Housefly** (B) Mango hopper

(C) Honey bee (D) Wasp

239. Dead heart and white head damage to rice is caused by

(A) Gall midge (B) Leaf roller

(C) Army worm (D) **Stem borer**

240. Wingless arthropods with 6 abdominal segments are found in which order?

(A) Diplura (B) Thysanura

(C) **Collembolan** (D) Protura

241. Legless grub is found in

(A) Meal moth (B) Khapra beetle

(C) **Rice weevil** (D) Lesser grain borer

242. Which term is used for the movement of coxa towards the body?
(A) Protraction (B) Promotion
(C) Adduction (D) **Abduction**

243. Which of the following structures is associated with articulation of wings?
(A) **Tegula** (B) Trochanter
(C) Coxa (D) Phragmata

244. Which of the following muscles are directly attached to the wing?
(A) Dorsal muscles (B) **Axillary muscles**
(C) Basilar muscles (D) Sub-alar muscles

245. Example of lateral sclerites is
(A) Basisternum (B) **Episternum**
(C) Pre-sternium (D) Sternum

246. Which of the following is an abdominal appendage?
(A) Cerci (B) Gonopore
(C) Waist (D) Apophysis

247. The terminal abdominal segments 6-9 are telescoped within the segment 5 when not in use. This condition is found in
(A) Gallfly (B) **Housefly**
(C) Dragonfly (D) Whitefly

248. The process of post- ecdysial hardening and darkening of cuticle is regulated by
(A) Ecdysone (B) PTTH
(C) Juvenile hormone (D) **Bursicon**

249. The ommatidia in which the samper cells produce hard crystalline cone are called
(A) Acone (B) **Eucone**
(C) Pseudocone (D) Exocone

250. Small apterous insects, compound eyes, reduced or absent ectoparasites of birds and mammals comes into
(A) Thysanura (B) Thysanoptera
(C) **Mallophaga** (D) Hymenoptera

251. How many pair of legs are found in mustard sawfly at larval stage?
(A) 4 pairs (B) 6 pairs
(C) **8 pairs** (D) 10 pairs

252. On which abdominal segments, legs of larval stage occurs?

(A) 3 to 5 and 10 (B) 4 to 8 and 10

(C) 2 to 4 and 10 (D) **3 to 6 and 10**

253. Monocondylic single mandible is found in

(A) Hemiptera (B) Hymenoptera

(C) **Thysanoptera** (D) Diptera

254. Insect having prolegs in larval stage

(A) ***Athalia proxima*** (B) *Aphis* spp.

(C) *Thrips tabaci* (D) *Dacus cucurbitae*

255. Example of order having pupal stage only in male

(A) Dictyoptera (B) Orthopteran

(C) Isopteran (D) **Strepsiptera**

256. Spiracles are absent in which insects?

(A) Cecidomyid larvae (B) ***Chironomid* larvae**

(C) Mycetophilid larvae (D) Psycodid larvae

257. In aphids, cornicles are located on

(A) 4th abdominal segment (B) **5th abdominal segment**

(C) 6th abdominal segment (D) 7th abdominal segment

258. Difference in male and female cockroach is

(A) Shape of the wing (B) Number of antennal segments

(C) Structure of compound eyes (D) **Presence or absence of anal style**

259. Stage of insects may be fully controlled

(A) **Egg** (B) Larva

(C) Pupa (D) Adult

260. Johnston"s organ is located in the

(A) Legs (B) wings

(C) Mouth parts (D) **2nd segment of antennae**

261. Number of segments found on insect head

(A) 4 (B) 5

(C) 7 (D) **6**

262. The antennae are appendages of which neuromere

(A) Tritocerebrum (B) **Deutocerebrum**

(C) Labial ganglion (D) Protocerebrum

263. Stadium is
   (A) The actual stage between two moult
   (B) **The time interval between two moult**
   (C) Both A&B
   (D) None of these

264. Instar is
   (A) **The actual stage between the two moults**
   (B) The time interval between the two moults
   (C) Both 'A' & 'B'
   (D) None of these

265. Adult mite have number of legs
   (A) 4 (B) 5
   (C) 6 **(D) 8**

266. Wings of insects are
   (A) Biangular (B) Quadriangular
   (C) Irregular shape (D) **Somewhat triangular**

267. One pair of wings is found in
   (A) Dictyoptera (B) Orthopteran
   (C) **Diptera** (D) Coleopteran

268. Degenerate ocelli of cockroach is known as
   (A) Cylassa (B) Acrotergite
   (C) **Fenestra** (D) None of these

269. Pterostigma is a character of order
   (A) **Odonata** (B) Orthoptera
   (C) Protura (D) Collembolan

270. Head and metathorax fused with each other is called
   (A) **Cephalothorax** (B) Pterothorax
   (C) Antennae (D) Mesopodeum

271. Stage of silk produce maximum silk
   (A) Prepupal stage (B) 4th instar
   (C) 5thinstar (D) **Both B&C**

272. Newly hatched silkworm is known as
   (A) Ants (B) **Kingo**
   (C) Chawaki (D) None of these

273. Nodus wings are present in order

(A) Orthoptera (B) **Odonata**

(C) Diptera (D) Coleopteran

274. Proboscis is modified part of

(A) Mandible (B) Maxillae

(C) **Labium** (D) Ypopharynx

275. Ligula is characteristic of

(A) Grasshopper (B) Beetle

(C) **Both A and B** (D) Dragonfly

276. Primarily wingless insects are found in which order

(A) **Coccids** (B) Odonata

(C) Protura (D) Orthoptera

277. Most important character of dictyoptera

(A) Elytra (B) Hemelytra

(C) **Large anal lobe in hind wing** (D) None of these

278. Exopterygota insect having pupal instar

(A) Odonata (B) Orthoptera

(C) Coleoptera (D) **Thysanoptera**

279. Example of secondry wingless insect is

(A) **Head louse** (B) Housefly

(C) Honey bee (D) None of the above

280. The composition of integuments is

(A) Cuticle membrane +basement membrane

(B) **Cuticle + epidermis +basement membrane**

(C) Cuticle + epidermis

(D) Epidermis + basement membrane

281. Anamorphosis is found in which order

(A) Orthoptera (B) Dictyoptera

(C) **Protura** (D) Coleoptera

282. Sting is modified part of

(A) Mouth part (B) Thorax

(C) Wings (D) **Ovipositor**

283. Non-reproductive type of abdominal appendages is

(A) Vagina & penis (B) **Cerci and prolegs**

(C) Anus & genital chamber (d) None of the above

284. Hook & channel type of wings coupling apparatus is found in

(A) **Diptera** (B) Lepidoptera

(C) Orthoptera (D) Coleoptera

285. Anal wing angle is found at

(A) Between apical & costal margin

(B) Between radius & medius vein

(C) **Between apical & anal margin**

(D) None of the above

286. Apical wing angle is found at

(A) Between costal & subcostal margin

(B) **Between costal & apical margin**

(C) Between radius & medius vein

(D) None of the above

287. Humeral wing angle is located at

(A) **Base of coastal vein** (B) Base of radius vein

(C) Base of medius vein (D) Base of subcostal vein

288. Medius vein is absent in which order

(A) Hymenoptera (B) **Hemiptera**

(C) Diptera (D) Thysanoptera

289. Hypothetical type of wing venation is proposed by

(A) **Comstock & Needham** (B) ADImms

(C) Tillyard (D) Srirastawa

290. In certain grasshopper, wings are short & this condition is known as

(A) Elytra (B) Hemelytra

(C) **Brachypterous** (D) Halteres

291. Which type of modification of wing is found in hemipterous insects

(A) Elytra (B) **Hemelytra**

(C) Brachypterous (D) Halteres

292. Which type of modification is found in strepsiptera

(A) **Pseudohalteres** (B) Elytra

(C) Brachypterous (D) Halteres

293. Suctorial & sucking type of legs occur in
(A) Male mosquito (B) **Male Dytiscus**
(C) Housefly (D) Male Honey bees

294. Climbing or sticking type of legs are found in
(A) Mosquito (B) Mustard sawfly
(C) **House fly** (D) Honey bees

295. Stridulatorial or sound producing legs are found in
(A) Forelegs of grasshopper (B) **Hind legs of male grasshopper**
(C) Hind wings of housefly (D) Forelegs of beetles

296. Which type of mouthparts is found in thrips
(A) Piercing & sucking type (B) Sponging type
(C) Siphoning type (D) **Rasping & sucking type**

297. Which type of mouthparts are found in honeybees
(A) Chewing & biting type (B) **Chewing & lapping type**
(C) Sponging type (D) Siphoning type

298. Capitate type antennae is found in
(A) **Coleoptera & Lepidoptera** (B) Homoptera & Heteroptera
(C) Orthoptera & Thysanoptera (D) Lepidoptera & Hymenoptera

299. Whorled type of antennae is found in
(A) Termites (B) Aphids
(C) **Male mango mealy bug** (D) Grasshopper

300. Function of epicuticle is
(A) **Prevent loss of water from insects integument**
(B) Prevent loss of nutrients from leaf surface
(C) Prevent loss of air from surface
(D) Prevent wax secretion

301. The term suture means
(A) Where two bones are attached (B) **Where two body regions are attached**
(C) Where two nerves are attached (D) All the above

302. The term Propodeum means
(A) Remaining part of abdomen (B) Waist of insects
(C) Parts of alimentary canal (D) **Fused segment of metathorax and abdomen**

303. The polymetabola metamorphosis is found in insect of family

(A) **Chalicididae** (B) Jassidae
(C) Fulgoridae (D) Pyralidae

304. The metamorphosis found in cockroach is

(A) Polymetabola (B) **Paurometabola**
(C) Hemimetabola (D) Holometabola

305. Halters which maintain balance during flight are present in

(A) Lepidoptera (B) **Diptera**
(C) Dermaptera (D) Hemiptera

367. The angle formed between the coastal and apical margin in insect wing is

(A) **Apical** (B) Marginal
(B) Humeral (D) Anal

368. The Johnston organ is present on

(A) Scape (B) **Pedicel**
(C) Coxa (D) Trochanter

369. The males are winged and females are wingless in

(A) Coccids (B) Embioptera
(C) Strepsiptera (D) **All of above**

370. In which of the following insect order, both pairs of wings are lost completely providing apterous condition

(A) Psocoptera (B) Mallophaga
(C) Siphonopter (D) **All of the above**

371. Only the basal part of wing is sclerotised forming

(A) Halteres (B) Elytra
(C) **Hemielytra** (D) Tegmina

372. The basal segment of leg is

(A) Trochanter (B) Pedicel
(C) **Coxa** (D) Tarsus

373. The sclerites present at wing base for joining thorax are collectively called as

(A) **Pteralia** (B) Fatigium
(C) Jegulae (D) None of the above

374. Which of the following is a true suture

(A) Epicranial (B) **Post occipital**

(C) Occipital (D) None of the above

375. In thrips , rasping and sucking type of mouthparts are assymetrical due to absence of

(A) Left mandible (B) **Right mandible**

(C) Maxillae (D) Labrum

376. In siphoning type of mouthparts watch spring type(proboscis) of structure is the modification of

(A) Maxillary palpi (B) Mandibles

(C) Stylets (D) **Galea**

377. Saltatorial legs are found in

(A) Cockroach (B) Preying mantis

(C) **Grasshopper** (D) Mole cricket

378. In coleopteran, dictyoptera, orthoptera, the forewing which appears to be extensively sclerotized is called

(A) Halteres (B) **Elytra and tegmina**

(C) Hemielytera (D) None of the above

379. Number of stylets in female mosquito and bugs is

(A) 2 & 6 (B) **4 & 6**

(C) 6 & 4 (D) 8 & 2

380. The wing coupling mechanism having small hooks to couple fore and hind wingstogether is

(A) Jugate (B) **Hamuli**

(C) Frenate (D) Amplexiform

381. Anamorphosis is encountered in which insect order

(A) **Protura** (B) Thysanura

(C) Collembola (D) Diplura

382. Propodeum is a segment of

(A) Head (B) **Abdomen**

(C) Thorax (D) Wings

383. Number of prolegs in looper, semilooper, caterpillar

(A) 4,6,7 (B) **2,3,5**

(C) 5,6,8 (D) 4,2,3

384. Example of exopterygote insect order with pupal instar in its life cycle
(A) Lepidoptera (B) Heteroptera
(C) Trichoptera (D) **Thysanoptera**

385. Primitive type of insect head
(A) **Hypognathous** (B) Prognathous
(C) Ophisthognathus (D) None of the above

386. The epistomal suture is absent in
(A) ***Blatta orientalis*** (B) Grasshopper
(C) Gryllids (D) Mantids

387. Physogastry occurs in
(A) **Termites** (B) Butterflies
(C) Diptera (C) Collembolans

388. Dolichasters is an abdominal appendage in
(A) **Antlion grub** (B) Moth larvae
(C) Odonata naids (D) Ephemeroptera nymphs

389. Cerci are absent in
(A) Hemiptera, odonata (B) Hemiptera,odonata, orthoptera
(C) **Hemiptera, Thysanoptera, Lepidoptera** (D) Orthoptera, odonata

390. Sridulatorial legs are present in
(A) **Hind legs of grasshoppers** (B) Forelegs of grasshoppers
(C) Forelegs of beetles (D) Middle legs of Grasshopper

391. Foragial or pollen collecting legs are present in
(A) **Worker honey bee** (B) Drones
(B) Queen bee (D) Soldiers

392. Clasping type legs are found in
(A) Sugarcane top borer (B) **Sugarcane leaf hopper**
(C) Sugarcane root borer (D) Sugarcane stem borer

393. Which type of wing modification is seen in strepsiptera
(A) **Pseudohalteres** (B) Elytra
(C) Brachypterous (D) Halters

394. Chrysalis type of pupa is seen in
(A) Houseflies (B) **Butterflies**
(C) Moths (D) Lacewings

395. Whorl type of antenna are seen in
(A) Hadda beetle (B) **Mealy bugs**
(C) Lac insect (D) Silkworm

396. The first Govt. entomologist of India is
(A) Stebbing (B) **Niceville**
(C) HS Pruthi (D) Ram Krishan Ayyar

397. Entomological Society of India Started in?
(A) **1938** (B) 1966
(C) 1972 (D) 1975

398. What is the major component of insect exoskeleton?
(A) Lipid (B) **Chitin**
(C) Protein (D) Cellulose

399. Which part of the exoskeleton lies between the exocuticle and the wax layer?
(A) Cuticulin layer (B) Endocuticle
(C) Procuticle (D) **Cement Layer**

400. Which part of the exoskeleton is composed of living cells?
(A) Basement membrane (B) Endocuticle
(C) Procuticle (D) **Epidermis**

401. Which is not a part of the Epicuticle
(A) Cuticulin layer (B) Wax layer
(C) **Exocuticle** (D) Cement layer

402. What is the function of wax layer
(A) Protection (B) Muscle attachment
(C) Thermal insulation (D) **Water conservation**

403. Rigid inflexible regions of the exoskeleton are called as
(A) Sutures (B) Apodemes
(C) **Sclerites** (D) Segments

404. A suture is best describes as
(A) **A junction between two sclerites**
(B) A thick ridge of cuticle
(C) A point of attachment between two segments
(D) A membranous bridge

405. Apodeme cannot be describe as

(A) **A point of attachment for muscles**

(B) A thick ridge of cuticle

(C) Invagination to strengthen the exoskeleton

(D) A flexible joint in the exoskeleton

406. Which of the following structure is unicellular

(A) Spine (B) **Seta**

(C) Spur (D) Microtrichia

407. Which of the following factors is most influencing for abundance of insects

(A) Functional wings (B) **Universal adaptability**

(C) High fecundity (D) Hexapod locomotion

408. The hard protein Sclerotin which gives horny nature is present in

(A) Endocuticle (B) **Exocuticle**

(C) Cement layer (D) Cuticulin

409. Chitin main component of Procuticle upon hydrolysis produces

(A) Citric Acid + H_2O (B) **Acetic acid + Glucosamine**

(C) Amino acid + Acetic Acid (D) Acetic Acid + Glutamic acid

410. Which among the following is a tanning hormone

(A) JH (B) MH

(C) **Bursicon** (D) Eclosion Hormone

411. Which among the following outgrowths of integuments has no articulations

(A) Setae (B) Spur

(C) **Bristles** (D) Spine

412. Odoriferous glands are present in

(A) Rice Plant Hopper (4th abdominal segment)

(B) Lac Insect

(C) **Rice Gundhi bug nymphs (5th abdominal segment)**

(D) RiceGundhi bug (3rd abdominal segment)

413. Which among the following is associated with moulting process

(A) Apolysis (B) Ecdysis

(C) Sclerotization (D) **All of the above**

414. The type of mimicry in which the mimic closely resembles an npalatable species and thus avoided by predator is

(A) Mullerian Mimicry (B) Crypsis

(C) **Batesian Mimicry** (D) Aposematism

415. The similarity in coloration between the monarch and viceroy butterflies is example of

(A) Crypsis (B) Mimesis

(C) Mullerian mimicry (D) **Batesian Mimicry**

416. Moulting fluid is able to digest

(A) Cement Layer (B) Cuticulin layer

(C) **Endocuticle** (D) Exocuticle

417. In the formation of cuticle which of the following layer is secreted first

(A) Exocuticle (B) Endocuticle

(C) Wax layer (D) **Cuticulin layer**

418. During the process of sclerotization which of the following layer is oxidised in the haemocytes

(A) **Arginine** (B) Glycine

(C) Tyrosine (D) Alanine

419. The one factor common to all arthropods and to which their success is largely attributed, is their

(A) Small size (B) Chitinous exoskeleton

(C) **Jointed appendage** (D) None of the above

420. Instar is

(A) **The actual stage between two moult**

(B) The time interval between two moult

(B) Both A and B

(C) None of above

421. Which of the following characters are not found in insects

(A) Segmented body (B) **Closed circulation**

(C) Jointed legs (D) Respiration through trachea

422. The Johnston organ is situated in which antennal segment

(A) First (B) **Second**

(C) Third (D) Fourth

423. Antennae are absent in

(A) Diptera (B) Coleoptera
(C) **Protura** (D) Hymenoptera

424. Bipectinate type of antennae are found in

(A) Grasshopper (B) Mustrard sawfly
(C) Bugs (D) **Silkworm Moths**

425. Adult mite possess

(A) Two pair of legs (B) Three pair of legs
(C) Four pair of legs (D) **Five pair of legs**

426. Ecdysone is secreted by

(A) Corpora Allata (B) **Prothoracic glands**
(C) Labial Glands (D) Mandibular Glands

427. Aristate type of antennae is found in

(A) Mosquito (B) **Housefly**
(C) Cockroach (D) Honeybee

428. Flabellate type of antennae is found in

(A) **Stylopid** (B) Honeybee
(C) Robber fly (D) Female mosquito

429. Whorled type of antennae is found in

(A) Termites (B) **Mango mealy bug**
(C) Aphids (D) Grasshopper

430. Clavate type of antennae is found in

(A) Click beetle (B) **Blister beetle**
(C) Dung roller (D) Red flour beetle

431. Geniculate type of antennae is found in

(A) Honeybee (B) Ants
(C) Weevils (D) **All of the above**

432. Smallest body part of insect is

(A) Leg (B) Head
(C) Thorax (D) Abdomen

433. Insect head is formed of how many segments

(A) 7 (B) **6**
(C) 4 (D) 5

434. Hypognathous head is found in

(A) Bugs (B) Beetles

(C) Weevils (D) **Grasshopper**

435. The long axis of head is horizontal and mouth parts point forward constitutes'

(A) Opisthognahous head (B) **Proganthous head**

(C) Hypognathous head (D) None of the above

436. Which is largest segment of antenna

(A) Pedicel (B) **Scape**

(C) Flagellomere (D) All of the above

437. The hind leg of insect are located on

(A) Prothorax (B) Mesothorax

(C) Abdomen (D) **Metathorax**

438. Insect legs are made up of how many components

(A) **5** (B) 3

(C) 4 (D) 4

439. Which is the largest part of insect leg

(A) Tibia (B) **Femur**

(C) Tarsus (D) Coxa

440. Lacinia is the part of

(A) Labrum (B) Mandible

(C) Labium (D) **Maxilla**

441. Halteres are found in

(A) **Diptera** (B) Lepidoptera

(C) Coleoptera (D) Hymenoptera

442. Tagmina is found in

(A) Diptera (B) Lepidoptera

(C) Coleoptera (D) **Orthoptera**

443. Ambulatorial type of legs are found in

(A) **Beetles** (B) Wasps

(C) Cockroaches (D) House fly

444. Scansorial type of insect legs are found in

(A) House fly (B) Honeybee

(C) Silkworm moth (D) **Head louse**

445. Stridulatorial /sound producing legs are found in

(A) **Forelegs of grasshopper** (B) Hind legs of male grasshopper

(C) Hindlegs of housefly (D) Forelegs of beetle

446. Which type of modification of wings is found in Orthoptera

(A) Elytra (B) Hemielytra

(C) **Tagmina** (D) Haltere

447. Which type of modification of wings is found in Coleoptera

(A) **Elytra** (B) Hemielytra

(C) Tagmina (D) Haltere

448. Hypothetical type of wings veination is proposed by

(A) **Comstock and Needham** (B) AD Imms

(C) Metcalf (D) Tillyard

449. Proboscis is modification part of

(A) Mandible (B) Maxillae

(C) **Labium** (D) Hypopharynx

450. Which of the following structure is associated with the articulation of the wings

(A) **Trochanter** (B) Tegula

(C) Scutum (D) Phragmata

451. Which of the following is the cross vein

(A) Costa (B) **Radial**

(C) Media (D) Cubitus

452. Berlese's theory was proposed to explain the origin of

(A) Egg stage (B) **Larval stage**

(C) Pupal Stage (D) Adult stage

453. Number of abdominal legs of larva of mustard sawfly

(A) 4 pairs (B) 6 pairs

(C) **8 pairs** (D) 10 pairs

454. Killing of silkworm pupa is known as

(A) Blowing (B) **Stifling**

(C) Denier (D) Hatching

455. Exarate pupa found in

(A) **Lepidoptera** (B) Coleoptera

(C) Diptera (D) Protura

456. Alula is situated at the base of
(A) Halteres (B) **Forewing of Diptera**
(C) Last Abdominal segmment (D) Forewing of Strepsiptera

457. The abdomen is specialized for
(A) Feeding and locomotion (B) Locomotion and reproduction
(C) **Reproduction and digestion** (D) Digestion and Ingestion

458. Following is example of Biopesticide
(A) DDT (B) BHC
(C) **NPV** (D) Malathion

459. How many abdominal segments are found in a typical insects
(A) Less than 5 (B) 6 to 11
(C) **12 or 13** (D) More than 13

460. The dorsal sclerite of each abdominal segment is called
(A) Sternite (B) Pleurite
(C) Coxite (D) **Tergite**

461. Which structure never occurs on an abdominal segment
(A) Spiracle (B) Spines
(C) Apodeme (D) Pleural structures

462. Junctions between the abdominal segments are best described as
(A) **Telescoping** (B) Dovetailed
(C) Ball and socket (D) Hinged

463. Which of the following structure is unpaired
(A) **Cerci** (B) Epiproct
(C) Valvifer (D) Paraproct

464. Which structure would never be found on an insect prothorax
(A) Legs (B) **Wings**
(C) Spiracle (D) Pronotum

465. To which body segment are the elytra attached
(A) **Mesothorax** (B) Metathorax
(C) Prothorax (D) First abdominal segment

466. What happens when you press down on then top of an insect thorax
(A) The wings go up (B) **The wings go down**
(C) The wings go forward (D) The wings go backward

467. Which of these is not a longitudinal wing vein

(A) Costa (B) Radius

(C) **Furca** (D) Cubitus

468. Axillary sclerites from points of attachments for muscles that control the

(A) Antennae (B) **Wings**

(C) Legs (D) Mouthparts

469. An insect head is specialized for

(A) Reproduction and digestion (B) Ingestion and precipitation

(C) **Orientation and locomotion** (D) Adaptation and respiration

470. The mouth parts include all of these structures except

(A) Hypopharynx (B) Maxillae

(C) **Clypeus** (D) Labrum

471. What is the name of the shell of an insect egg

(A) **Chorion** (B) Morula

(C) Cerosa (D) Blastula

472. The main characteristic features of insects is

(A) Pair of antennae (B) **Three pair of legs**

(C) Presence of wings (D) All the above

473. The damaging stage of whit grub is

(A) **Grubs and adult** (B) Larva and adult

(C) Larva only (D) Only adult

474. The function of an acropyle is

(A) **Prevent water loss** (B) Absorb moisture

(C) Gaseous exchange (D) Excrete nitrogenous waste

475. Which of these occupies the largest volume in an unfertilized insect egg

(A) Periplasm (B) Yolk

(C) **Nucleus** (D) Embryo

476. Which layer of exoskeleton forms after ecdysis has been completed

(A) Procuticle (B) **Exocuticle**

(C) Epicuticle (D) Cuticular layer

477. Which larval type does not have walking legs

(A) Scarabaeiform (B) **Vermiform**

(C) Eruciform (D) Campodeiform

478. Which structure could be found in both males and females

(A) **Valvifer** (B) Aedeagus

(C) Epiproct (D) Clasper

479. What is the maximum number of ocelli that may be found in an adult insect

(A) Zero (B) **Three**

(C) Five (D) Twenty

480. Which mouth arts bear palps

(A) Labrum and Labium (B) Labium and Maxillae

(C) **Maxillae and Mandibles** (D) Mandibles and Labrum

481. In order to be more effective control measures aimed at killing the insect should be directed to attack

(A) Eggs stage (B) **Pupal stage**

(C) Larval stage (D) Adult stage

482. Metamorphosis include

(A) Change in nymph (B) Changes in behaviour

(C) **Changes from egg to adult stage**

(D) Mandibles and Labrum

483. The type of legs in grasshopper

(A) Raptorial type (B) Fossorial type

(C) Cursorial type (D) **Saltatorial type**

484. Entomology is the branch of

(A) **Zoology** (B) Botany

(C) Chemistry (D) Agriculture

485. Piercing –sucking type of mouth parts are found in

(A) Butterflies (B) Thrips

(C) **Mosquitoes** (D) Beetles

486. The term Entomology is derived from

(A) **Greek word** (B) French word

(C) German word (D) English word

487. Rasping and sucking type of mouth parts are found in

(A) **Thrips** (B) Termites

(C) Locust (D) Beetles

Unit-2

Insect Physiology

1. The Father of Insect physiology is
   (A) L. I. Gilbert (B) **V. B. Wigglesworth**
   (C) G. A. Kerkut (D) R. F. Chapman

2. The author of book "Insect Physiology" is
   (A) L. I. Gilbert (B) **V. B. Wigglesworth**
   (C) G. A. Kerkut (D) R. F. Chapman

3. V. B. Wigglesworth conducted experiments with which insect
   (A) ***Rhodnius prolixus*** (B) Blow fly
   (C) Cecropia moth (D) Locust

4. Caroll Williams conducted experiments with which insect
   (A) *Rhodnius prolixus* (B) Blow fly
   (C) **Cecropia moth** (D) Locust

5. Morris Rockstein conducted experiments with which insect
   (A) *Rhodnius prolixus* (B) ***Phormia***
   (C) Cecropia moth (D) Locust

6. Regnald F. Chapman conducted experiments with which insect
   (A) *Rhodnius prolixus* (B) *Phormia*
   (C) Cecropia moth (D) ***Locusta migratoria***

7. Vince Gaston Dethier conducted experiments with which insect
   (A) *Rhodnius prolixus* (B) ***Phormia regina***
   (C) Cecropia moth (D) *Locusta migratoria*

8. The author of book "Principles of Insect Physiology" is
   (A) L. I. Gilbert (B) **V. B. Wigglesworth**
   (C) G. A. Kerkut (D) R. F. Chapman

9. The author of book " Physiology of Insecta " is

(A) G.A. Kertkut and L. I. Gilbert (B) V. B. Wigglesworth

(C) **M. Rockstein** (D) R. F. Chapman

10. The author of book "Insect Physiology and Anatomy" is

(A) L. I. Gilbert (B) V. B. Wigglesworth

(C) G. A. Kerkut (D) **N.C. Pant and S. Ghai**

11. The Indian scientist who studied nutritional requirements of insects and the role of symbionts is

(A) H. S. Pruthi (B) S. Ghai

(C) S. Pradhan (D) **N.C. Pant**

12. As per the Dyar's Law, the width of the head capsule increased in a regular linear progression in successive instars by a ratio of

(A) **1.3-1.7** (B) 2.3-2.7

(C) 3.3-3.7 (D) 4.3-4.7

13. As per, Przibram's Law of growth, all liner dimensions of insect during each instar are increased by a ratio of

(A) **2.3** (B) **1.26**

(C) 3.26 (D) 4.39

14. As per, Przibram's Law of growth, an insect's weight during each instar is

(A) **Doubled** (B) Quadruple

(C) Three times (D) Six times

15. The foregut is also known as

(A) **Stomodaeum** (B) Mesentron

(C) Proctodaeum (D) Ventriculus

16. The Mid gut is also known as

(A) Ventriculus (B) Mesentron

(C) **Both A and B** (D) Proctodaeum

17. The Hindgut is also known as

(A) Stomodaeum (B) Mesentron

(C) **Proctodaeum** (D) Ventriculus

18. The Functional stomach is

(A) Stomodaeum (B) **Mesentron**

(C) Proctodaeum (D) All of the above

19. The circular valve between stomodaeum and mesenteron is known as

(A) Stomodael valve (B) Cardiac valve

(C) **Both A and B** (D) Pyloric valve

20. The valve between mesenteron and proctodaem is known as

(A) Proctodael valve (B) Pyloric valve

(C) **Both A and B** (D) Cardiac valve

21. The Cibarial pump is well developed in

(A) **Sucking insect** (B) biting and chewing insect

(C) Both A and B (D) None of the above

22. The origin of stomodaeum is

(A) **Ectodermal** (B) Endodermal

(C) Mesodermal (D) All the above

23. The cavity where pre oral cavity open is known as

(A) Salivarium (B) **Cibarium**

(C) Pharynx (D) None of theabove

24. The cavity where salivary glands open is known as

(A) **Salivarium** (B) Cibarium

(C) Pharynx (D) None of theabove

25. The number of muscular ridges present in proventriculus/gizzard is

(A) 4 (B) **6**

(C) 8 (D) 10

26. The proventriculus/gizzard is absent in

(A) Cockroach (B) Grasshopper

(C) Beetles (D) **Aphids**

27. The pH of midgut is

(A) **Alkaline** (B) Acidic

(C) Neutral D) None the above

28. The range of midgut pH is

(A) 4-6 (B) 7

(C) **6-8** (D) 6-7

29. The pH of midgut in Lepidoptera and trichoptera is

(A) 4-6 (B) 6-8

(C) **8-10** (D) 6-7

30. The origin of midgut is
    (A) Ectodermal (B) **Endodermal**
    (C) Mesodermal (D) All the above

31. The peritrophic membrane is present in
    (A) Stomodaeum (B) **Mesentron**
    (C) Proctodaeum (D) All of the above

32. The function of peritrophic membrane is
    (A) Help in digestion of food
    (B) **Permit passage of food particle for digestion**
    (C) Assimilation of food particle
    (D) Secretion of enzymes

33. The peritrophic membrane is
    (A) Permeable (B) **Semi permeable**
    (C) Impermeable (D) None of the above

34. The number of Gastric cecae present in grasshopper is
    (A) 4 (B) **6**
    (C) 8 (D) 10

35. The function of gastric cecae is
    (A) Help in digestion of food
    (B) Permit passage of food particle for digestion
    (C) Assimilation of food particle
    (D) **Increase surface area of mid git for absorption**

36. The Filter chamber is present in
    (A) **Aphids** (B) Grasshopper
    (C) Cockroach (D) Termites

37. The filter chamber is formed
    (A) **The posterior part of the hindgut is in contact with the anterior part of the foregut**
    (B) The Anterior part of the hindgut is in contact with the anterior part of the foregut
    (C) The posterior part of the hindgut is in contact with the Posterior part of the foregut
    (D) None of the above

38. The salivary glands in caddisfly is modified for

(A) Toxin production (B) **Silk production**

(C) Both A and B (D) Not modifed

39. The salivary glands in Honeybee is modified for

(A) Toxin production (B) Silk production

(C) **Hormone production** (D) Anticoagulant production

40. The salivary glands in Mosquitoes is modified for

(A) Toxin production (B) Silk production

(C) Hormone production (D) **Anticoagulant production**

41. The water is reabsorbed in

(A) **Rectum** (B) Colon

(C) midgut (D) Ileum

42. The wool and feathers are digested with the help of which enzyme

(A) Pectinase (B) Collagenase

(C) Aminopeptidaes (D) **Keratinase**

43. The rectal glands function as respiratory organ is

(A) Mecoptera (B) **Odonata**

(C) Dictyoptera (D) Trichoptera

44. The water is reabsorbed in

(A) **Rectum** (B) Colon

(C) midgut (D) Ileum

45. A diet with all components of known quantity is referred to as

(A) Meridic diet (B) Oligidic diet

(C) **Holidic diet** (D) Artificial diet

46. A diet with whose components are not fully knwon is referred to as

(A) Meridic diet (B) **Oligidic diet**

(C) Holidicdiet (D) Artificial diet

47. A diet with some components of known quantity is referred to as

(A) **Meridic diet** (B) Oligidic diet

(C) Holidic diet (D) Artificial diet

48. A diet with host plant material + supplement nutrients is referred to as

(A) Meridic diet (B) Oligidic diet

(C) Holidic diet (D) **Xenic diet**

49. The endosymbionts are present in
(A) **10-15% insects** (B) 15-30% insects
(C) 30-45% insects (D) >50 insects

50. The fungus Garden termites are
(A) ***Macrotermes* sp.** (B) *Odontotermes* sp.
(C) *Coptotemes* sp. (D) All of the above

51. The endosymbionts present in termites are
(A) **Flagellate protozoa** (B) Bacteria
(C) Fungus (D) All of the above

52. The Malphigian tubules was discovered by
(A) Meckel (B) **Marcello Malphigi**
(C) Wigglesworth (D) None of the above

53. The Malphigian tubules was discovered by Marcello Malphigi in the year
(A) **1669** (B) 1769
(C) 1869 (D) 1829

54. The Malphigian tubules was named by
(A) **Meckel** (B) Marcello Malphigi
(C) Wigglesworth (D) None of the above

55. The origin of Malphigian tubules is
(A) Endodermal (B) **Ectodermal**
(C) Mesodermal (D) Odonata

56. The Malphigian tubules are absent in
(A) Hempitera (B) Thysanura
(C) **Collembola** (D) Odonata

57. The Malphigian Tubule are vestigial in
(A) Protura (B) Strepsiptera
(C) Diplura (D) **All of the above**

58. The number of malphigian tubules in Coccids is
(A) **2** (B) 10
(C) 24 (D) 200

59. The number of malphigian tubules in Hemiptera and Thysanura is
(A) **4** (B) 10
(C) 24 (D) 200

60. The number of malphigian tubules in Odonata and Orthoptera is

(A) 4 (B) 10

(C) 24 (D) **200**

61. The excretory product of insect is mostly

(A) Ammonia (B) **Uric acid**

(C) Urea (D) None of these

62. The insects are in general

(A) Ammonotelic (B) **Uricotelic**

(C) Ureotelic (D) None of these

63. The aquatic insects are

(A) **Ammonotelic** (B) Uricotelic

(C) Ureotelic (D) None of these

64. The insects in water limited areas is

(A) Ammonotelic (B) **Uricotelic**

(C) Ureotelic (D) None of these

65. The advantage of being uricotelic is

(A) Less toxic (B) Water insoluble

(C) Low H:Nratio **(D) All of the above**

66. The distal 2/3rd part of malphigian tubules consist of

(A) **Honey comb cells** (B) Brush bordered cells

(C) Urate (D) None of these

67. The Proximal 1/3rd part of malphigian tubules consist of

(A) Honey comb cells (B) **Brush bordered cells**

(C) Urate (D) None of these

68. The distal 2/3rd part of malphigian tubules is concerned with the

(A) Absorption of water (B) Absorption of minerals

(C) **Both A and B** (D) None of these

69. The Proximal 1/3rd part of malphigian tubules is concerned with the

(A) Re-absorption of water (B) Re-absorption of minerals

(C) **Both A and B** (D) None of these

70. The condition where MT are long and lying free in the body cavity

(A) **Fundamental type** (B) Coleopterous type

(C) Hemipteran type (D) Lepidopterous type

71. The Primitive or fundamental type of MT are present in
    (A) **Orthoptera** (B) Coleoptera
    (C) Hemiptera (D) Lepidoptera

72. The condition where MT are attached distally to one side of the gut or may be disposed of around the rectum in a simple convulated layer
    (A) Fundamental type (B) **Coleopterous type**
    (C) Hemipteran type (D) Lepidopterous type

73. The condition where MT are attached distally to one side of the gut or may be disposed of around the rectum in a simple convulated layer
    (A) Cryptonephridial type (B) Coleopterous type
    (C) **Both A and B** (D) Lepidopterous type

74. The Cryptonephridial type of MT are present in
    (A) Orthoptera (B) **Coleoptera**
    (C) Hemiptera (D) Lepidoptera

75. The condition where MT are free at distal ends but the proximal half is opaque and the distal half is clear
    (A) Fundamentaltype (B) Coleopterous type
    (C) **Hemipteran type** (D) Lepidopterous type

76. The condition where MT are Cryptonephridial type but the proximal half is opaque and the distal half is clear
    (A) Fundamental type (B) Coleopterous type
    (C) Hemipteran type (D) **Lepidopterous type**

77. The Storage excretion is present in
    (A) **Collembola** (B) Thysanura
    (C) Orthoptera (D) Dictyoptera

78. The Labial Kidney is present in
    (A) Collembola (B) Thysanura
    (C) **Both A and B** (D) None of the above

79. The opening of Labial Kidney is
    (A) **At the base of Labium** (B) At 9th abdominal segment
    (C) At 6th abdominal segment (D) At the prothorax

80. The Nephrocytes are present in
    (A) Butterflies (B) *Glossina*
    (C) **Honeybee** (D) Cockroach

81. The urine production in insects is regulated by which hormone
(A) **Diuretic Hormone** (B) Anti-Diuretic Hormone
(C) Bursicon (D) PTTH

82. Which of the following play role in excretion in insects
(A) Labial glands (B) Nephrocytes
(C) Pericardial cells (D) **All of the above**

83. The circulatory system in insects
(A) **Open type** (B) Closed type
(C) Intermediate type (D) None of the above

84. Put the odd one out
(A) Perivisceral sinus (B) Pericardial sinus
(C) Perineural cells (D) **Dorsal blood vessel**

85. The fibro-muscular septum between Pericardial and Perivisceral sinus is known as
(A) **Dorsal Diaphragm** (B) Ventral Diaphragm
(C) Heart (D) Dorsal blood vessel

86. The fibro-muscular septum between Perivisceral and Perineural sinus is known as
(A) Dorsal Diaphragm (B) **Ventral Diaphragm**
(C) Heart (D) Dorsal blood vessel

87. The body cavity of where the haemolymph freely flow is known as
(A) Perivisceral sinus (B) **Haemocoel**
(C) Heart (D) Dorsal blood vessel

88. The circulatory system of insect consists of
(A) Dorsal blood vessel (B) Heart
(C) Aorta (D) **All of the above**

89. The Ostia are present in
(A) Dorsal blood vessel (B) **Heart**
(C) Aorta (D) All of the above

90. The Ostia are Absent in
(A) Dorsal blood vessel (B) Heart
(C) **Aorta** (D) All of the above

91. The Fan shaped muscles with which heart remain attached with the tergum are
(A) **Alary muscles** (B) Cardial muscles
(C) Pericardial Muscles (D) All of the above

92. The primitive number of Alary muscles in Orthoptera is
(A) 6 (B) **12**
(C) 13 (D) 14

93. The chamber of heart in cockroach is
(A) 6 (B) 10
(C) **13** (D) 4

94. The chamber of heart in Honeybee is
(A) 6 (B) 10
(C) 13 (D) **4**

95. The chamber of heart in House fly is
(A) **3** (B) 6
(C) 10 (D) 13

96. The chamber of heart in Chironomid larva is
(A) **2** (B) 6
(C) 10 (D) 13

97. The Aorta is Pulsatile in
(A) Chironomids (B) ***Dytiscus* sp.**
(C) *Periplenata* sp. (D) All of the above

98. The primitive number of ostia in insects is
(A) 2 (B) 6
(C) **12** (D) 13

99. The number of ostia in most of the lepidopteran insects
(A) 2-4 (B) **7-8**
(C) 10-12 (D) 4-6

100. The number of ostia in Collembola
(A) 2-4 (B) **5-6**
(C) 10-12 (D) 4-6

101. The number of ostia in House fly
(A) 2 (B) **4**
(C) 6 (D) 8

102. The ostia allowing haemolymph in to dorsal blood vessel
   (A) **Incurrent Ostia** (B) Excurrent Ostia
   (C) Alary Ostia (D) Cardial Ostia

103. The ostia allowing haemolymph out of dorsal blood vessel
   (A) Incurrent Ostia (B) **Excurrent Ostia**
   (C) Alary Ostia (D) Cardial Ostia

104. The organs located near the base of the antennae, wings or legs helping in blood circulation are known as
   (A) **Accessory Pulsatile Organs** (B) Accessory Cardial Organs
   (C) Accessory Pericardial Organs (D) All of the above

105. The rate of heartbeat in cockroach is
   (A) 29/min (B) **49/min**
   (C) 79/min (D) 99/min

106. The rate of heartbeat in *Hippobasca* sp. is
   (A) 29/min (B) 49/min
   (C) 90/min (D) **120/min**

107. The blood sugar is
   (A) Glucose (B) Galactose
   (C) Mannose (D) **Trehalose**

108. The anti freezing component in haemolymph is
   (A) Glucose (B) Galactose
   (C) **Glycerol** (D) Trehalose

109. The Green insectoviridin pigment is present in
   (A) Scale insect (B) Aphids
   (C) Bombyx larva (D) **Locusts**

110. The haemolymph or blood in insect lack
   (A) Vitamin E (B) **Vitamin K**
   (C) Vitamin C (D) Vitamin D

111. The Purplish red protaphin pigment is present in
   (A) Scale insect (B) **Aphids**
   (C) Bombyx larva (D) Locusts

112. The blood pH is

(A) **6-7** (B) 7-8

(C) 8-9 (D) 5-6

113. The most abundant haemocytes in blood is

(A) **Plasmatocytes** (B) Granulocytes

(C) Oenocytes (D) Spherule cells

114. The haemocytes number in insects in general is

(A) 10,000-20,000mm^3 (B) **20,000-40,000mm^3**

(C) 30,000-50,000mm^3 (D) 40,000-50,000mm^3

115. The pigment Haemoglobin is present in

(A) *Chironomus* (B) *Gasterophilus*

(C) *Anisops* (D) **All the above**

116. The carrier of Oxygen in insect is

(A) Haemolymph (B) Haemocytes

(C) **Trachea** (D) All the above

117. The origin of fat body is

(A) Ectodermal (B) **Mesodremal**

(C) Endodermal (D) All the above

118. The function of fat bodies is the synthesis and secretion of

(A) Vitellogenin (B) MFOs

(C) Diapause proteins (D) **All the above**

119. The number of neurons in insects is approximately

(A) 1 lakh (B) **5 Lakh**

(C) 10 Lakh (D) 15 Lakh

120. The basic unit of nervous system is

(A) **Neuron** (B) Ganglia

(C) Axon (D) Brain

121. The neuron did not consists of

(A) Cell body (B) Axon

(C) Dendrite (D) **Synapse**

122. The neurons having a single stalk from the cell body that connects with the axon

(A) **Unipolar neuron** (B) Multipolar neuron

(C) Bipolar neuron (D) Mixed neuron

123. The neurons of the cell body bears an axon and a single, branched dendrite

(A) Unipolar neuron (B) Multipolar neuron

(C) **Bipolar neuron** (D) Mixed neuron

124.The neurons having an axon and several branched dendrites

(A) Unipolarneuron (B) **Multipolar neuron**

(C) Bipolar neuron (D) Mixed neuron

125. The nerves which pick up stimulus and generate an action potential

(A) Interneuron (B) Neuroendocrine cell

(C) **Sensory nerve** (D) Mixed nerve

126.The nerves which effect a response, e.g., gland releases or muscle movement

(A) Interneuron (B) Neuroendocrine cell

(C) Sensory nerve (D) **Mixed nerve**

127. The nerve to nerve communication is mediated by

(A) **Interneuron** (B) Neuroendocrine cell

(C) Sensorynerve (D) Mixed nerve

128. The nerves cells which secrete hormone

(A) Interneuron (B) **Neuroendocrine cell**

(C) Sensory nerve (D) Mixed nerve

129. The nervous system in insects is

(A) Centralized (B) **Decentralized**

(C) Autonomous (D) All of the above

130. Embryologically, the number of ganglia in insect nervous system s

(A) 12 (B) 15

(C) 18 (D) **20**

131. The Ganglia within each segment are linked to one another by a short medial nerve called

(A) Interneuron (B) Neuroendocrine cell

(C) Connective (D) **Commissures**

132. In cyclorrhapous dipterans, the fusion of c*arpora cardiaca, carpora allata* and prothoracic glands is refereed to as

(A) Craber's organ (B) Simpher's organ

(C) **Weismann'sring** (D) Gyroscopic organ

133. The Ganglia within each segment are linked to ganglia in adjacent body segment by a short medial nerve called

(A) Interneuron (B) Neuroendocrine cell

(C) **Connectives** (D) Commissures

134. The brain of insect is formed by the fusion of

(A) 3 ganglia (B) 5 ganglia

(C) **6 ganglia** (D) 8 ganglia

135. The hypocerebaral or supra oesophageal ganglia is formed by the fusion of

(A) Protocerebrum (B) Deutocerebrum

(C) Deutocerebrum (D) **All the above**

136. The Protocerebral ganglia is concerned with

(A) Vision (B) Neuro-endocrine control of moulting

(C) Smell and learning (D) **All the above**

137. The Deutocerebral ganglia is concerned with

(A) Antennal function (B) Innervate olfatcory lobe

(C) Innervate lateral Ocelli (D) **All the above**

138. The Suboesophageal ganglia is formed by the fusion of

(A) Mandibular Ganglion (B) Maxillary ganglion

(C) Labial ganglion (D) **All the above**

139. The most coomon neurotransmitter in insect nervous system is

(A) **Acetylcholine** (B) Dopamine

(C) GABA (D) All the above

140. The Prothoracotropic Hormone (PTTH) is secreted by

(A) **Prothoracic glands** (B) Corpora allata

(C) Neuro secretory cells (D) All the above

141. The Molting hormone or Ecdysone is secreted by

(A) **Prothoracic glands** (B) Corpora allata

(C) Neuro secretory cells (D) All the above

142. The Juvenile hormone or Ecdysone is secreted by

(A) Prothoracicglands (B) **Corpora allata**

(C) Neuro secretory cells (D) All the above

143. The tracheal system is absent is some

(A) Collembola (B) Protura

(C) Endoparasitic larvae (D) **All the above**

144. The origin of trachea is

(A) **Ectodermal** (B) Mesodermal

(C) Endodermal (D) All the above

145. The ring like annulations along the trachea is

(A) Spiracle (B) **Taenidia**

(C) Intima (D) Atrium

146. The taenidia is absent in

(A) **Air sac** (B) Primary trachea

(C) Secondary trachea (D) Secondary trachea

147. The opening of trachea on integument is

(A) **Spiracle** (B) Taenidia

(C) Intima (D) Atrium

148. The smallest branches of Trachea is

(A) Spiracle (B) Taenidia

(C) **Tracheole** (D) Atrium

149. The cavity with valve like opening over spiracle is

(A) Spiracle (B) Taenidia

(C) Intima (D) **Atrium**

150. The main longitudinal trunk of trachea is formed by

(A) Dorsal tracheal trunk (B) Visceral tracheal trunk

(C) ventral tracheal trunk (D) **All the above**

151. The number of spiracles in Diplura is

(A) 6 pairs (B) 8 pairs

(C) 10 pairs (D) **11 pairs**

152. In general, the number of spiracles in insects is

(A) **10 (2 pairs on thorax and 8 pairs on abdomen)**

(B) 11 (3 pairs on thorax and 8 pairs on abdomen)

(C) 12 (3 pairs on thorax and 9 pairs on abdomen)

(D) 13 (3 pairs on thorax and 10 pairs on abdomen)

153. The number of spiracles in Hover flies and syrphid flies is

(A) **2 pairs** (B) 4 pairs

(C) 6 pairs (D) 8 pairs

154. The number of spiracles in Mosquito larvae is

(A) **1 pairs** (B) 2 pairs
(C) 5 pairs (D) 8 pairs

155. The spiracles are not interconnected in

(A) **Collembola** (B) Protura
(C) Diplura (D) Thysanura

156. The spiracle having 1 mesothoracic + 1 metathoracic + 8 abdominal spiracles functional is

(A) **Holopneustic type** (B) Peripneustic type
(C) Hemipneustic type (D) Oligopneustic type

157. The Holopneustic type tracheal system is present in

(A) ***Bibionid* larva** (B) Ceccidomyid larva
(C) Psychodid larva (D) Mycetophilid larva

158. The spiracle having 1 mesothoracic + 1 metathoracic + 7 abdominal spiracles functional is

(A) Holopneustic type (B) **Peripneustic type**
(C) Hemipneustic type (D) Oligopneustic type

159. The Peripneustic type tracheal system is present in

(A) *Bibionid* larva (B) **Ceccidomyid larva**
(C) Psychodid larva (D) Mycetophilid larva

160. The spiracle having 1 mesothoracic + 7 abdominal spiracles functional is

(A) Holopneustic type (B) Peripneustic type
(C) **Hemipneustic type** (D) Oligopneustic type

161. The Hemipneustic type tracheal system is present in

(A) Bibionid larva (B) Ceccidomyid larva
(C) Psychodid larva (D) **Mycetophilid larva**

162. The spiracle having one or two spiracles functional is

(A) Holopneustictype (B) Peripneustic type
(C) hemipneustic type (D) **Oligopneustic type**

163. The spiracle having 1 mesothoracic + 1 Post abdominal spiracle functional is

(A) **Amphipneustic type** (B) Metapneustic type
(C) Propneustic type (D) Oligopneustic type

164. The Amphipneustic type tracheal system is present in

(A) *Bibionid* larva (B) Ceccidomyid larva

(C) **Psychodid larva** (D) Mycetophilid larva

165. The spiracle having only mesothoracic spiracle functional is

(A) Amphipneustic type (B) Metapneustic type

(C) **Propneustic type** (D) Oligopneustic type

166. The Propneustic type tracheal system is present in

(A) **Dipterous larva** (B) Ceccidomyid larva

(C) Psychodid larva (D) Mycetophilid larva

167. The spiracle having Post abdominal spiracle functional is

(A) Amphipneustic type (B) **Metapneustic type**

(C) Propneustic type (D) Oligopneustic type

168. The Metapneustic type tracheal system is present in

(A) *Bibionid* larva (B) *Culicid* larva

(C) Psychodid larva (D) Mycetophilid larva

169. The spiracle having 1 mesothoracic + 1 Post abdominal spiracle functional is

(A) **Amphipneustic type** (B) Metapneustic type

(C) Propneustic type (D) Oligopneustic type

170. The condition where none of the spiracle is functional

(A) **Apneustic type** (B) Metapneustic type

(C) Propneustic type (D) Oligopneustic type

171. The Apneustic condition is found in

(A) Aquatic insects (B) Endo parasites

(C) **Both A and B** (D) None of the above

172. In insects having Apneustic type of tracheal system, exchange of gases is

(A) Blood (B) **Integument**

(C) Both A and B (D) None of the above

173. The insects without full complements of ten spiracles, but where one or more pairs disappear completely are

(A) Apneustic type (B) **Hypopneustic type**

(C) Both A and B (D) All the above

174. The Hypopneustic condition is found in

(A) Thrips (B) Scale insects

(C) **Both A and B** (D) None of the above

175. The volume of O_2 used by insect at rest

(A) **0.6-3 litres of O_2/kh/hr** (B) 6-10 litres of O_2/kh/hr

(C) 10-20 litres of O_2/kh/hr (D) None of the above

176. The ventilation mechanism in insects is

(A) Active (B) Passive

(C) **Both A and B** (D) None of the above

177. The concentration of oxygen in water

(A) **4.9%** (B) 10.9%

(C) 15.9% (D) 21.9%

178. The Blood gills are present in

(A) Trichoptera (B) Diptera

(C) **Both A and B** (D) None of the above

179. The Rectal gills are present in

(A) Trichoptera (B) Diptera

(C) **Odonata** (D) None of the above

180. The Hydrofuge hairs are present in

(A) Plecoptera (B) **Diptera**

(C) Trichoptera (D) None of the above

181. An extensive network of fine tracheal beneath the integument, either concentrated in some regions or distributed all over aquatic insects

(A) **Tracheal gills** (B) Spiracular gills

(C) Physical gills (D) None of the above

182. The Tracheal gills are present in

(A) Plecoptera (B) Ephemeroptera

(C) Trichoptera (D) **All of the above**

183. The spiracular gills are present in

(A) Coleoptera (B) Diptera

(C) **Both A and B** (D) None of the above

184. The Respiratory siphon (Breathing Tubes) are present in

(A) *Mansonia* sp. (B) Syrphid fly larva

(C) Water scorpion (D) **All of the above**

185. The Plastron act as a
(A) Tracheal gill (B) Spiracular gill
(C) **Physical gill** (D) None of the above

186. The insects having plastron respiration are
(A) Flood water mosquitoes (B) Riffle beetles
(C) **Both A and B** (D) None of the above

187. The respiratory pigment Haemoglobin is present in
(A) *Buenoa* sp. (B) *Gasterpohilus* sp.
(C) *Anisops* sp. (D) **All of the above**

188. The insects having plastron respiration are
(A) Flood water mosquitoes (B) Riffle beetles
(C) **Both A and B** (D) None of the above

189. The Respiratory quotient is a measure of
(A) Respiration rate (B) **Metabolism**
(C) Both A and B (D) None of the above

190. If the Respiratory quotient value =1, it indicate the metabolism of
(A) Proteins (B) **Carbohydrate**
(C) Lipids (D) None of the above

191. If the Respiratory quotient value =0.7, it indicate the metabolism of
(A) Proteins (B) Carbohydrate
(C) **Lipids** (D) None of the above

192. The function of Neurosecretory cells (NSC) is
(A) Produce Brain hormone (B) Produce Diuretic hormone
(C) Produce Bursicon (D) **All of the above**

193. The Hormone inducing Urine formation in insects is
(A) Produce Brain (B) **Diuretic hormone**
(C) Bursicon (D) All of the above

194. The hormone stimulating thoracic glands to secrete Moulting Hormone is
(A) **Brain hormone** (B) Diuretic hormone
(C) Bursicon (D) All of the above

195. The hormone facilitating darkening and hardening of cuticle of adult insects is
(A) Brain hormone (B) Diuretic hormone
(C) **Bursicon** (D) All of the above

196. The Neuro-secretion from the NSC of the brain are stored temporarily in

(A) Corpora Allata (B) **Corpora Cardaica**
(C) Prothoracic Glands (D) None of the above

197. The Weismann's Ring is present in

(A) Odonata (B) **Diptera**
(C) Lepidoptera (D) Orthoptera

198. The Juvenile Hormone/Neotenin is secreted by

(A) **Corpora Allata** (B) Corpora Cardaica
(C) Prothoracic Glands (D) None of the above

199. The Thoracic Glands was First described by

(A) Buetnandt and Karlson (B) **Lyonet**
(C) Karl Salama and Caroll Williams (D) None of the above

200. The Moulting Hormone Ecdysone was First isolated by

(A) **Buetnandt and Karlson (1954)** (B) Lyonet
(D) None of the above (C) Karl Salama and Caroll Williams

201. The Primer pheromone affect

(A) **Growth and Development** (B) Behaviour
(C) Both A and B (D) None of the above

202. The Releaser pheromone affect

(A) Growth and Development (B) **Behaviour**
(C) Both A and B (D) None of the above

203. The Moulting Hormone Ecdysone was First isolated by

(A) **Buetnandt and Karlson (1954)** (B) Lyonet
(C) Karl Salama and Caroll Williams (D) None of the above

204. The First Sex Pheromone, Bombykol was isolated by

(A) **Buetnandt *et al.* (1959)** (B) Lyonet
(C) Karl Salama and Caroll Williams (D) None of the above

205. The ductless glands are called as

(A) **Endocrine glands** (B) Exocrine Glands
(C) Ductless Glands (D) None of the above

206. The Frontal glands are present in
   (A) Stink bug (B) **Termites**
   (C) Slug Caterpillar (D) Honeybee

207. The Setal glands are present in
   (A) **Stink bug** (B) Termites
   (C) Slug Caterpillar (D) Honeybee

208. The Mandinular glands are present in
   (A) Stinkbug (B) Termites
   (C) Slug Caterpillar (D) **Honeybee**

209. The Repugnatorial glands are present in
   (A) **Stink bug** (B) Termites
   (C) Slug Caterpillar (D) **Honeybee**

210. The Androconia (Scented glands) are present in
   (A) Citrus butterfly larva (B) **Pieris brassica Males**
   (C) Slug Caterpillar (D) Honeybee

211. The Osmetaria (Forked glands) are present in
   (A) **Citrus butterfly larva** (B) *Pieris brassica*
   (C) Slug Caterpillar (D) Honeybee

212. The Interspecific semiochemicals include
   (A) Synomones (B) Kairomones
   (C) Allomones (D) **All of the above**

213. The **Intraspecific semiochemicals include**
   (A) Sex Pheromone (B) Alarm Pheromone
   (C) AggregationPheromon (D) **All of the above**

214. The **Interspecific** semiochemical that favours the producers
   (A) Synomones (B) Kairomones
   (C) **Allomones** (D) All of the above

215. The Interspecific semiochemical that favours the recievers
   (A) Synomones (B) **Kairomones**
   (C) Allomones (D) All of the above

216. The **Interspecific** semiochemical that favours both producers and receivers
   (A) **Synomones** (B) Kairomones
   (C) Allomones (D) All of the above

217. The External fertilization is present in
(A) Silverfish (B) Bristletails
(C) Collembola (D) **All of the above**

218. The male couplatory organ in insect is called as
(A) Vasa efferntia (B) **Aedeagus**
(C) Spematids (D) None of the above

219. The process of sperm formation in insect is
(A) Spermatogonia>Spermatids>Spermatocytes> Spermatozoa
(B) **Spermatogonia>Spermatocytes>Spermatids>Spermatozoa**
(C) Spermatogonia>Spermatocytes>Spermatozoa >Spermatids
(D) Spermatids>Spermatogonia>Spermatocytes> Spermatozoa

220. The Male Spematogonia are
(A) Haploid (B) **Diploid**
(C) Haplo diploid (D) Diplo haploid

221. The Male Spematids are
(A) **Haploid** (B) Diploid
(C) Haplo diploid (D) Diplo haploid

222. The accessory glands in male reproductive system help in
(A) Manufacture of **seminalfluid** (B) Production of **spermatophores**
(C) **Both A and B** (D) None of the above

223. The paired penis is present in
(A) **Ephemeroptera** (B) Odonata
(C) Phasmatodea (D) Grylloblatodea

224. The male reproductive system is present in
(A) 8th Abdominal segment (B) **9thAbdominal segment**
(C) 10thAbdominal segment (D) 11th Abdominal segment

225. The female insect ovaries with no special nutritive/nurse/trophocytes cells
(A) Meroistic type (B) **Panoistictype**
(C) Acrotrophic type (D) Polytrophic type

226. The Panoistic type ovaries are present in
(A) Ephemeroptera (B) Thysanura
(C) Thysanoptera (D) **All of the above**

227. The female insect ovaries with special nutritive/nurse/trophocytes cells

(A) Dermaptera (B) Hemiptera

(C) Lepidoptera (D) **All of the above**

228. The Meroistic type ovaries are present in

(A) Ephemeroptera (B) Thysanura

(C) Thysanoptera (D) **All of the above**

229. The female insect ovaries where trophocytes accompany each oocyte into vitellarium

(A) Meroistictype (B) Panoistictype

(C) Acrotrophic type (D) **Polytrophictype**

230. The Polytrophic type ovaries are present in

(A) Hymenoptera (B) Lepidoptera

(C) Diptera (D) **All of the above**

231. The female insect ovaries where trophocytes remain in germarium connection to oocyte via nutritive cord

(A) Meroistictype (B) Panoistictype

(C) **Acrotrophictype** (D) Polytrophic type

232. The Acrotrophic/Telotrophic type ovaries are present in

(A) Coleoptera (B) Hemiptera

(C) **Both A and B** (D) None of the above

233. The suborder Adephaga of Coleoptera have

(A) Meroistic ovaries (B) Panoistictype ovaries

(C) Acrotrophic type ovaries (D) **Polytrophic ovaries**

234. The suborder Polyphaga of Coleoptera have

(A) Meroistic type ovaries (B) Panoistic type ovaries

(C) **Acrotrophic type ovaries** (D) Polytrophic type ovaries

235. Both Polytrophic and Panoistic ovaries are present in

(A) Siphonoptera (B) Neuroptera

(C) Mecoptera (D) **All of the above**

236. The Parental care is exhibited in insects

(A) *Gryllotalpa* (B) Bees

(C) Termites (D) **All of the above**

237. The Queen substance in honeybee is
(A) 9-oxo-trans-2-decenoic acid (B) 9-ODA
(C) **Both And B** (D) All of the above

238. The reproduction where eggs are laid shortly after fertilization is
(A) **Oviparity** (B) Ovoviviparity
(C) Viviparity (D) None of the above

239. The reproduction where eggs retained until embryogenesis complete and are laid immediately before or soon after hatching
(A) Oviparity (B) **Ovoviviparity**
(C) Viviparity (D) None of the above

240. The reproduction where eggs retained and Embryo fed by mother and the embryo completes development inside the egg while inside the mother and comes out as larva or nymph
(A) Oviparity (B) Ovoviviparity
(C) **Viviparity** (D) None of the above

241. The Larviparous viviparity is present in
(A) ***Sarcophaga sp.*** (B) Aphids
(C) Tsetse fly (D) All of the above

242. The Nymphiparous viviparity is present in
(A) *Sarcophaga* sp. (B) **Aphids**
(C) Tsetse fly (D) All of the above

243. The Pupiparous viviparity is present in
(A) *Sarcophaga* sp. (B) Aphids
(C) **Tsetse fly** (D) All of the above

244. The adenotrophic viviparity is present in
(A) *Hippobasca variegata* (B) *Glossina palpalis*
(C) **Both A and B** (D) *Mycophila speyeri*

245. Inadenotrophic viviparity, larvae feeds in
(A) Haemocoel (B) **Uterus**
(C) Placenta (D) None of the above

246. The Haemocoelic viviparity is present in
(A) *Gall midges* (B) *Mycophila speyeri*
(C) Strepsiptera (D) **All of the above**

247. The Pseudoplacental viviparity is present in

(A) *Hemimermis talpoides* (B) *Macrosiphum* sp.
(C) Psocoptera (D) **All of the above**

248. The Reproduction by immature or larval stage is termed as

(A) **Paedogenesis** (B) Gynogenesis
(C) Parthenogenesis (D) Pseudogamy

249. The phenomenon of Paedogenesis is observed in

(A) *Henria* sp. (B) *Miastor metralaos*
(C) *Micromalthus debilis* (D) **All of the above**

250. The asexual reproduction where two or more eggs are produced from a single egg is known as

(A) **Polyembrony** (B) Gynogenesis
(C) Hermaphroditism (D) Pseudogamy

251. The phenomenon of Polyembryony is observed in

(A) *Platygaster* sp. (B) *Bracon* sp.
(C) *Aphelopus* sp. (D) **All of the above**

252. The phenomenon of Hermaphroditism is observed in

(A) *Icerya purchasii* (B) *Icerya bimaculata*
(C) *Icerya zeteki* (D) **All of the above**

253. The parthenogenetic development of the egg requiring the sperm to initiate the development but without fusing with the female nucleus is known as

(A) Gynogenesis (B) Pseudogamy
(C) **Both A and B** (D) Functional Hermaphroditism

254. The phenomenon of Gynogenesis/Pseugamy is observed in

(A) Spider beetle (B) Psychid moth
(C) Bark beetle (D) **All of the above**

255. The Polyploidy is recorded in

(A) **Silkworm** (B) Psychid moth
(C) Bark beetle (D) All of the above

256. The insects displaying Caste polymorphism are

(A) Honeybees (B) Termites
(C) Ants (D) **All of the above**

257. The Cyclic parthenogenesis is also known as

(A) Gynogenesis (B) Pseudogamy

(C) **Heterogony** (D) Functional Hermaphroditism

258.The Cyclic parthenogenesis/Heterogony is observed in

(A) Jassids (B) **Aphids**

(C) Scales (D) Miaster

259. The parthenogenesis where only females are produced

(A) **Thelytoky** (B) Arrhenotoky

(C) Amphitoky (D) Heterogony

260. The Thelytokus parthenogenesis is observed in

(A) *Oecophylla longinoda* (B) *Apis* sp.

(C) **Thysanura** (D) Aphids

261. The parthenogenesis where only males are produced

(A) Thelytoky (B) **Arrhenotoky**

(C) Amphitoky (D) Heterogony

262. The Arrhenotokus parthenogenesis is observed in

(A) *Oecophylla longinoda* (B) ***Apis* sp.**

(C) Thysanura (D) Aphids

263. The parthenogenesis where both male and females are produced

(A) Thelytoky (B) Arrhenotoky

(C) **Amphitoky** (D) Heterogony

264. The Amphitokus parthenogenesis is observed in

(A) ***Oecophylla longinoda*** (B) *Apis* sp.

(C) Thysanura (D) Aphids

265.The Obligatory parthenogenetic insect is

(A) *Warramba virgo* (B) *Chelonus blackburni*

(C) **Both And B** (D) All of the above

266. The occasional/accidental parthenogenesis is known as

(A) Apomictic parthenogenesis (B) Arrhenotoky

(C) **Tychoparthenogenesis** (D) Heterogony

267. The basic unit of the muscle is

(A) **Sarcomere** (B) Sarcolemma

(C) Both A and B (D) None of the above

268. The type of muscles found in insects
(A) Cardiac (B) Smooth muscle
(C) **Striated** (D) All of the above

269. The Neuro modulators of muscle activity are
(A) Sertonin (B) Octopamine
(C) Proctolin (D) **All of the above**

270. The Inhibitory Neuro tarsmiiter of muscle activity is
(A) Sertonin (B) Octopamine
(C) Proctolin (D) **GABA**

271. The wing movement and most of the flight is controlled by
(A) **Direct flight muscles** (B) Indirect flight muscles
(C) Both A and B (D) None of the above

272. The Sliding filament theory of muscles was given by
(A) **Huxley *et al.*** (B) Regnald F. Chapman
(C) V.B. Wigglesworth (D) All of the above

273. The Queen substance in honeybee is
(A) 9-oxo-trans-2-decenoic acid (B) 9-ODA
(C) **Both And B** (D) All of the above

Unit-3

Insect Systematics

1. The Father of Taxonomy is
   (A) De Candolle (B) **Linnaeus**
   (C) Aristotle (D) Michael Anderson

2. The Father of Biological classification is
   (A) De Candolle (B) Linnaeus
   (C) **Aristotle** (D) Michael Anderson

3. The Father of Numerical Taxonomy is
   (A) De Candolle (B) Linnaeus
   (C) Aristotle (D) **Michael Anderson**

4. The term Taxonomy was coined by
   (A) **De Candolle** (B) Linnaeus
   (C) Aristotle (D) Michael Anderson

5. The Binomial Nomenclature was introduced by
   (A) De Candolle (B) **Linnaeus**
   (C) Aristotle (D) Michael Anderson

6. The Systema Naturae is written by
   (A) De Candolle (B) **Linnaeus**
   (C) Fisher (D) A. D. Imms

7. The First edition of Systema Naturae was published in
   (A) 1858 (B) 1835
   (C) 1758 (D) **1735**

8. The 10th edition of Systema Naturae was published in
   (A) 1858 (B) 1835
   (C) **1758** (D) 1735

9. The Binominal Nomenclature in Systema Naturae was published in
   - (A) 1st Edition (B) 7th Edition
   - (C) 9th Edition (D) **10th Edition**

10. The word Taxonomy is derived from
   - (A) Italic (B) Latin
   - (C) **Greek** (D) Arabic

11. The word Nomenclature is derived from
   - (A) Italic (B) **Latin**
   - (C) **Greek** (D) Arabic

12. The word Systematic is derived from
   - (A) Italic (B) **Latin**
   - (C) Greek (D) Arabic

13. The naming of organisms and groups of organism, rule and procedures to be followed in such naming is known as
   - (A) Taxonomy (B) Systemtics
   - (C) Classification (D) **Nomenclature**

14. The theoretical study of classification, including its basis, principles, procedure and rules is known as
   - (A) **Taxonomy** (B) Systemtics
   - (C) Classification (D) None of the above

15. The science of diversity of organisms
   - (A) Taxonomy (B) **Systemtics**
   - (C) Classification (D) None of the above

16. The taxonomy concerned with naming and characterization of species is
   - (A) **Alpha Taxonomy** (B) Beta Taxonomy
   - (C) Gamma Taxonomy (D) None of the above

17. The taxonomy concerned with Classification is
   - (A) Alpha Taxonomy (B) **Beta Taxonomy**
   - (C) Gamma Taxonomy (D) None of the above

18. The taxonomy concerned with evolutionary relations and phylogeny
   - (A) Alpha Taxonomy (B) Beta Taxonomy
   - (C) **Gamma Taxonomy** (D) None of the above

19. Ordering of organisms in to groups or sets on the basis of their relationships
    (A) Taxonomy (B) Systemtics
    (C) **Classification** (D) None of the above
20. The Classification help to
    (A) Identification of species
    (B) To trace phylogenetic relationship
    (C) Provide system of information retrieval
    (D) **All the above**
21. The basic unit of taxonomic classification is
    (A) **Species** (B) Genus
    (C) Family (D) Class
22. The rules regarding the naming of organisms is governed by
    (A) **ICZN** (B) ZSI
    (C) NHS (D) All of the above
23. The ICZN secretariat is located at
    (A) USA (B) **UK**
    (C) Canada (D) Italy
24. The underlying principle in the ICZN is the
    (A) Law of superiority (B) **Law of Priority**
    (C) Law of Preference (D) All of the above
25. The single specimen selected by the author of a species as its type, or the only specimen known at the time of description is
    (A) Paratype (B) Allotype
    (C) **Holotype** (D) All of the above
26. If holotype is damaged irreparably or lost, the specimen selected by the authority in group to replace is called
    (A) Paratype (B) Allotype
    (C) Holotype (D) **Neotype**
27. A specimen cited in the original description other than the holotype.
    (A) **Paratype** (B) Allotype
    (C) Holotype (D) Neotype
28. A specimen of the opposite sex to the type.
    (A) Paratype (B) **Allotype**
    (C) Holotype (D) Neotype

29. The underlying principle in the ICZN is the
    (A) Law of superiority (B) **Law of Priority**
    (C) Law of Preference (D) All of the above

30. If the author name is put in bracket after scientific name, it means
    (A) The taxa is originally described by the authorityin same group
    (B) **The taxa has been reclassified and placed in another group**
    (C) Both A and B
    (D) All of the above

31. A group of real organisms recognised as a formal unit at any level of hierarchic classification is
    (A) **Taxon** (B) Genus
    (C) Family (D) Class

32. Groups of interbreeding natural populations that are reproductively isolated from other such groups
    (A) **Species** (B) Genus
    (C) Family (D) Class

33. The superfamily has suffix
    (A) -idae (B) **-oidea**
    (C) -ini (D) None of the above

34. The family has suffix
    (A) **-idae** (B) -oidea
    (C) -ini (D) None of the above

35. The subfamily has suffix
    (A) **-idae** (B) -oidea
    (C) -ini (D) **-inae**

36. The Tribe has suffix
    (A) -idae (B) -oidea
    (C) **-ini** (D) None of the above

37. The basic unit of taxonomic classification is
    (A) **Species** (B) Genus
    (C) Family (D) Class

38. The group of real organisms recognised as a formal unit at any level of hierarchic classification.
    (A) Species (B) **Taxon**
    (C) Genus (D) Family

39. Group of interbreeding natural populations that are reproductively isolated from other such groups.
    (A) **Species** (B) Taxon
    (C) Genus (D) Family

40. The class insecta was divided in to Apterygota and Pterygota by
    (A) Linneaus (B) Michael Anderson
    (C) Aristotle (D) **Brawer**

41. Antennae are absent in
    (A) Crustacea (B) **Arachnida**
    (C) Onychophora (D) Myriapoda

42. Gill are found in
    (A) **Crustacea** (B) Arachnida
    (C) Onychophora (D) Myriapoda

43. Millipedes belong to
    (A) **Diplopoda** (B) Chilopoda
    (C) Arachnida (D) Onychophora

44. Prosoma, mesosoma and Metasoma are characteristic of
    (A) Chilopoda (B) **Arachnida**
    (C) Crustacea (D) Insecta

45. Four pair of locomotor are found in
    (A) Crustacea (B) **Arachnida**
    (C) Onychophora (D) Myriapoda

46. Antennae are absent in
    (A) Crustacea (B) **Arachnida**
    (C) Onychophora (D) Myriapoda

47. Vasiform orifice is a characteristic feature of
    (A) Delphacidae (B) **Aleyrodidae**
    (C) Membracidae (D) Cercopidae

48. Which one of the following is the largest order in insecta
(A) **Coleoptera** (B) Lepidoptera
(C) Homoptera (D) Hymenoptera

49. Classification of insects by Handlirsch was based on
(A) Wing morphology (B) Nutrition
(C) **Fossil forms** (D) Reproduction

50. Trigamma in lepidopteran wing is a structure formed by the veins
(A) M3, A1, Cu1b (B) M3, Cu1b, Cu2
(C) M3, Cu1a, A3 (D) **M3, Cu1a, Cu1b**

51. The insect orders which are not recorded from India are
(A) Zoraptera and Trichoptera (B) **Grylloblattodea and Zoraptera**
(C) Embioptera and Zoraptera (D) Grylloblattodea and Strepsiptera

52. Gamma taxonomy refers to
(A) Study of speciation (B) Species characterization
(C) Species identification (D) **Study of evolution**

53. Pregenital abdominal appendages are present in
(A) Lepidoptera (B) **Collembola**
(C) Coleoptera (D) Hymenoptera

54. Cerci are asymmetrical in
(A) **Male Embids** (B) Female Embids
(C) Female Cockroach (D) Male Cockroach

55. Halteres are present in
(A) **Diptera** (B) Coccidae
(C) Aleyrodidae (D) Jassidae

56. Anamorphosis is a character of
(A) **Protura** (B) Diplura
(C) Collembola (D) Thysanura

57. Book louse belong to order
(A) Protura (B) **Psocoptera**
(C) Thysanura (D) Zoraptera

58. Silk, which is secreted by silkworm contains
(A) Glutamin (B) Keratin
(C) Collagen (D) **Fibrion and Sericin**

59. Which among the following orders are NOT placed in panarpoid complex

(A) **Thysanoptera** (B) Neuroptera
(C) Lepidoptera (D) Mecoptera

60. Book lungs are characteristic of

(A) Onychophora (B) **Arachnida**
(C) Myriapoda (D) Crustacea

61. Worm like un-segmented body is found

(A) **Onychophora** (B) Arachnida
(C) Myriapoda (D) Crustacea

62. Paired stalked compound eyes are found in

(A) Onychophora (B) Arachnida
(C) Myriapoda (D) **Crustacea**

63. One pair of locomotor organs in each segment in present in

(A) **Chilopoda** (B) Diplopoda
(C) Onychophora (D) Myriapoda

64. Millipedes are

(A) **Herbivorous** (B) Carnivorous
(C) Both Herbivore and carnivorous (D) Feed on organic matter

65. Woodlice belong to

(A) Onychophora (B) Arachnida
(C) Myriapoda (D) **Crustacea**

66. Ticks and mites belong to

(A) Onychophora (B) **Arachnida**
(C) Myriapoda (D) Crustacea

67. *Peripatus* sp. belong to

(A) **Onychophora** (B) Arachnida
(C) Myriapoda (D) Crustacea

68. Mandibular articulation in apterygotes is

(A) Monocondylic (B) **Dicondylic**
(C) Absent (D) None of the above

69. Mandibular articulation in Pterygotes is

(A) **Monocondylic** (B) Dicondylic
(C) Absent (D) None of the above

70. Pregenital abdominal appendages are absent in

(A) **Apterygote** (B) Pterygote
(C) Endopterygote (D) Exopterygote

71. Pregenital abdominal appendages are present in

(A) Apterygote (B) **Pterygote**
(C) Endopterygote (D) Exopterygote

72. The newly created insect order is

(A) Anoplura (B) Phasmatodea
(C) **Mantophasmatodea** (D) Mantodea

73. Snakeflies, Alderflies belong to

(A) Lepidoptera (B) **Neuroptera**
(C) Mecoptera (D) Diptera

74. Stoneflies belong to

(A) **Plecoptera** (B) Neuroptera
(C) Mecoptera (D) Diptera

75. Scorpionflies belong to

(A) Lepidoptera (B) Neuroptera
(C) **Mecoptera** (D) Diptera

76.Stylopids belong to

(A) **Strepsiptera** (B) Neuroptera
(C) Mecoptera (D) Diptera

77. The insect order in Hemipteroid compelx

(A) Mallophaga (B) Hemiptera
(C) Siphunculata (D) **All of the above**

78. The insect order in Paleopteran complex

(A) Ephemeroptera (B) Odonata
(C) **Both A and B** (D) Plecoptera

79. Pupal stage is absent or rarely present in

(A) Pterygota (B) **Exopterygota**
(C) Endopterygota (D) None of the above

80. The immature stages of Exopterygotes resemble adults in

(A) **Structure and habit** (B) Structure
(C) Habit (D) Differ completely

81. The wings develop as wing buds in
    (A) Pterygota (B) Exopterygota
    (C) **Endopterygota** (D) None of the above

82. The wings develop as wing pads in
    (A) Pterygota (B) **Exopterygota**
    (C) Endopterygota (D) None of the above

83. The Bristletails belong to
    (A) Protura (B) Collembola
    (C) **Thysanura** (D) Diplura

84. Gladiator belong to order
    (A) Anoplura (B) Phasmatodea
    (C) **Mantophasmatodea** (D) Mantodea

85. Caddisflies belong to order
    (A) Mecoptera (B) **Trichoptera**
    (C) Strepsiptera (D) Neuroptera

86. Lacewing belong to order
    (A) Mecoptera (B) Trichoptera
    (C) Strepsiptera (D) **Neuroptera**

87. Scorpionflies belong to order
    (A) **Mecoptera** (B) Trichoptera
    (C) Strepsiptera (D) Neuroptera

88. Stylopids belong to order
    (A) Mecoptera (B) Trichoptera
    (C) **Strepsiptera** (D) Neuroptera

89. Fleas belong to order
    (A) Mecoptera (B) Trichoptera
    (C) **Siphonoptera** (D) Neuroptera

90. True flies belong to order
    (A) **Diptera** (B) Lepidoptera
    (C) Hymenoptera (D) Neuroptera

91. Webspinners belong to order
    (A) Mecoptera (B) Trichoptera
    (C) Strepsiptera (D) **Embioptera**

92. 6-segmented abdomen is characteristic of
   (A) **Collembola** (B) Protura
   (C) Diplura (D) Thysanura

93. In Collembola, the jumping organ is known as
   (A) Hamula (B) **Furcula**
   (C) Collophore (D) Retinaculum

94. In Collembola, the jumping organ (Furcula) is present on
   (A) 1st abdominal segment (B) 3rd abdominal segment
   (C) **4th abdominal segment** (D) 5rd abdominal segment

95. In Collembola, the Retinaculum/Hamula or Catch is present on
   (A) 1st abdominal segment (B) **3rd abdominal segment**
   (C) 4th abdominal segment (D) 5rd abdominal segment

96. In Collembola, the Collophore or Glue bar is present on
   (A) **1st abdominal segment** (B) 3rd abdominal segment
   (C) 4th abdominal segment (D) 5th abdominal segment

97. The Collembola as pest of Alfalfa is
   (A) ***Sminthurus viridis*** (B) *Hypogastrura armata*
   (C) Fire brats (D) Japygids

98. The appendages present in Protura
   (A) Antennae (B) Eyes
   (C) Cerci (D) **All**

99. The Protura is characterised by
   (A) Epimorphosis (B) **Anamorphosis**
   (C) Hypermetamorphosis (D) metamorphosis

100. The successive addition of segments in protura is between
   (A) 4th segment and telson (B) 6th segment and telson
   (C) **8th segment and telson** (D) 10th segment and telson

101. The appendage used as sensory organ/anetnnae in Protura
   (A) **1st pair of legs** (B) 2nd pair of legs
   (C) 3rd pair of legs (D) All three of legs

102. The example of Thysanura is
   (A) Silverfish (B) Firebrats
   (C) Bristle tail (D) **All of the above**

103. The silverfish feeding on wallpaper paste, book binding *etc.*

(A) *Lepisma sacharina* (B) *Ctenolepisma* sp.

(C) **Both A and B** (D) *Japyx* sp.

104. Mayflies, Dayflies, shadflies belong to order

(A) **Ephemeroptera** (B) Mecoptera

(C) Embioptera (D) Trichoptera

105. The order where adult insect have atrophied mouthparts

(A) **Ephemeroptera** (B) Mecoptera

(C) Embioptera (D) Trichoptera

106. Sub imago is characteristic of

(A) **Ephemeroptera** (B) Mecoptera

(C) Embioptera (D) Trichoptera

107. The pterygote order moults after becoming adult

(A) **Ephemeroptera** (B) Mecoptera

(C) Embioptera (D) Trichoptera

108. Tegmina is characteristic of

(A) Grasshopper (B) Crickets

(C) Gryllids (D) **All of the above**

109. Grasshopper lay eggs as

(A) **Egg Pod** (B) Egg mass

(C) Egg case (D) Egg batch

110. The suborder of Orthoptera having antenna segment >30

(A) Caelifera (B) **Ensifera**

(C) Gryllidae (D) Acrididae

111. The suborder of Orthoptera having antenna segment <30

(A) **Caelifera** (B) Ensifera

(C) Gryllidae (D) Acrididae

112. In Long horned grasshoppers (Ensifera), tympanal organs are present on

(A) **Fore tibia** (B) Either side of 1st abdominal segment

(C) Hind tibia (D) Either side of 3rd abdominal segment

113. In Short horned grasshoppers (Caelifera), tympanal organs are present on

(A) Foretibia (B) **Either side of 1st abdominal segment**

(C) Hind tibia (D) Either side of 3rd abdominal segment

114. The example of Long horned grasshoppers (Ensifera)
(A) Katydid (B) House cricket
(C) Mole cricket (D) **All of the above**

115. The Rice grasshopper belong to family
(A) Gryllotalpidae (B) **Acrididae**
(C) Gryllidae (D) Tettigonidae

116. The suborder of Orthoptera having antenna segment <30
(A) **Caelifera** (B) Ensifera
(C) Gryllidae (D) Tettigonidae

117. Locusts are
(A) **Grasshopper** (B) Katydids
(C) Crickets (D) Locusts

118. All locusts are grasshoppers. But not all grasshoppers locusts.
(A) **I^st^ statement is true but 2nd is False**
(B) I^st^ statement is False but 2nd is True
(C) Both the statements are true
(D) Both the statements are False

119. The multiplication of grasshoppers into locusts under favourable conditions represents
(A) Biological transformation (B) Phase chage
(C) **Ecological opportunism** (D) None of the above

120. Phase theory of locust was given by
(A) **B. P. Uvarov** (B) Y. R. Rao
(C) S. Pradhan (D) None of the above

121. The two phase theory proposed by Uvarov include
(A) Phasis solitaria and Phasis Transiens
(B) **Phasis solitaria and Phasis Gregaria**
(C) Phasis Gregaria and Phasis Transiens
(D) None of the above

122. The characteristic of Solitary phase (Phasis Solitaria) of Locust is
(A) 27-30 segmented antennae (B) Solitary feeder
(C) Longer and crested pronotum (D) **All of the above**

123. The characteristic of Gregaroius phase (Phasis Gregaria) of Locust is

(A) 27 segmented antennae (B) Solitary feeder

(C) Shorter and saddle pronotum (D) **All of the above**

124. The India Scientist having authority on grasshopper/locust

(A) M. S. Mani (B) **Y. R. Rao**

(C) S. Pradhan (D) None of the above

125. The most destructive locust is

(A) *Patanga succincta* (B) *Schistocerca gregaria*

(C) ***Locusta migratoria*** (D) *Nomadacris septemfasciata*

126. The scientific name of Bombay locust is

(A) ***Patanga succincta*** (B) *Schistocerca gregaria*

(C) *Locusta migratoria* (D) *Nomadacris septemfasciata*

127. The scientific name of Desert locust is

(A) *Patanga succincta* (B) ***Schistocerca gregaria***

(C) *Locusta migratoria* (D) *Nomadacris septemfasciata*

128. The scientific name of African Red locust is

(A) *Patanga succincta* (B) *Schistocerca gregaria*

(C) *Locusta migratoria* (D) ***Nomadacris septemfasciata***

129. The Locust Warning Organisation was established in

(A) **1939** (B) 1945

(C) 1951 (D) 1965

130. The Headquarter of Locust Warning Organisation is at

(A) Ajmer (B) Raipur

(C) **Jodhpur** (D) Jaipur

131. Phase theory of locust was given by

(A) **B. P. Uvarov** (B) Y. R. Rao

(C) S. Pradhan (D) None of the above

132. The order Phasmida includes

(A) Stick insect (B) Lea insect

(C) **Both A and B** (D) All of the above

133. The ability to regenerate the lost appendage is known as

(A) **Autotomy** (B) Apoptosis

(C) Thanatosis (D) All of the above

134. The Autotomy is characteristic of insect order

(A) Orthoptera (B) Mallophaga

(C) **Phasmida** (D) Strepsiptera

135. The Mimicry/crypsis is characteristic of insect order

(A) Orthoptera (B) Mallophaga

(C) **Phasmida** (D) Strepsiptera

136. The unsegmented, sclerotised Forcep like cerci is characteristic of insect order

(A) Orthoptera (B) Mallophaga

(C) Phasmida (D) **Dermaptera**

137. The Parental care is exhibited in insect order

(A) Orthoptera (B) Mallophaga

(C) Phasmida (D) **Embioptera**

138. The silk glands in order Embioptera are present in

(A) **Basitarsus of fore legs** (B) Basitarsus of Middle legs

(C) Basitarsus of hind legs (D) Basitarsus of All three pair of legs

139. The Order Odonata includes

(A) Snake Doctors (B) Devil's Darning Needles

(C) Mosquito Hawk (D) **All of the above**

140. The Order Odonata lacks the ability to fold wings when rest, they belong to

(A) Neoptera group (B) **Paleopteragroup**

(C) Hemipteroid group (D) None of the above

141. Immature stages of Odonata are

(A) **Aquatic** (B) Aerial

(C) Subterranean (D) None of the above

142. Immature stages of Odonata are specifically known as

(A) Nymphs (B) **Naiads**

(C) Caterpillar (D) Larvae

143. In Naiads, the mouthparts are modified as

(A) Catcher (B) **Mask**

(C) Gobbler (D) None of the above

144. Naiads are

(A) Herbivorous (B) **Predators**

(C) Omnivorous (D) None of the above

145. In Naiads, the which part of mouthparts is modified as Mask

(A) Mandible (B) Maxillae

(C) **Labium** (D) Labrum

146. The typical Pterostigma in Order Odonata is formed due to the thickening of wing membrane between

(A) **Costa and Radius Vein** (B) Radius and Medius Vein

(C) Costal and Radius vein (D) Medius and Cubitus vein

147. The Cerci in Odonata is

(A) **One segmented** (B) 3 segmented

(C) Multi segmented (D) 5 segmented

148. In Odonata, the functional couplatory organ in male is present on

(A) First abdominal segment (B) **Second abdominal segment**

(C) Eighth abdominal segment (D) Ninth abdominal segment

149. Dragonflies belong to sub order

(A) **Anisoptera** (B) Zygoptera

(C) Mecoptera (D) None of the above

150. Damselflies belong to sub order

(A) Anisoptera (B) **Zygoptera**

(C) Mecoptera (D) None of the above

151. The type of eyes present in Dragonflies is

(A) **Holoptic** (B) Dichoptic

(C) Both Holoptic and Dichoptic (D) None of the above

152. The type of eyes present in Damsonflies is

(A) Holoptic (B) **Dichoptic**

(C) Both Holoptic and Dichoptic (D) None of the above

153. The naiads of Dragonflies have

(A) **Rectal gills** (B) Caudal gills

(C) Tracheal gills (D) None of the above

154. The Cockroach and praying mantis belong to order

(A) Odonata (B) **Dictyoptera**

(C) Isoptera (D) Thysanoptera

155. The eggs in case of cockroaches are contained in

(A) **Ootheca** (B) Egg mass

(C) Egg case (D) None of the above

156. The number of eggs in ootheca of cockroach is

(A) 08 (B) **16**

(C) 32 (D) 46

157. The insect having 3600 vision is

(A) Grasshopper (B) **Praying mantids**

(C) Cockroach (D) Honeybee

158. The 'Y' shaped epicranial suture is typical of which insect order

(A) Anoplura (B) Isoptera

(C) **Psocoptera** (D) Mallophaga

159. In Psocoptera, which part of the maaxillae is the modified as 'Pick'

(A) **Lacinia** (B) Galea

(C) Cardo (D) Stipe

160. The polymorphic castes of termites are

(A) Supplemetary caste (B) Reproductive caste

(C) Sterile Caste (D) **All of the above**

161. The type of antennae in Isoptera is

(A) Geniculate (B) **Moniliform**

(C) Serrate (D) Capitate

162. The wings in termites generally break off and shed from

(A) **Humeral suture** (B) Jugate suture

(C) Radial suture (D) Costal suture

163. The remanant behind after wing shedding at humeral suture in termites is called

(A) Scar (B) **Scale**

(C) Stump (D) All of the above

164. The Termite species where humeral suture is absent and wings are irregularly torn off

(A) *Odontotermes* (B) ***Mastotermes***

(C) *Macrotermes* (D) All of the above

165. The Physogastry is characteristic of

(A) **Female (queen) termite** (B) Male (King) termite

(C) Wokers (D) Soldiers

166. The phenomenon of Physogastry is found in family

(A) **Termitidae** (B) Kalotermitidae

(C) Hetertermitidae (D) **All of the above**

167. In general the queen termite lay

(A) **One egg/second** (B) 1500 eggs/day

(C) 500 eggs/day (D) 10,000 eggs /day

168. The soldiers having median rostrum and vestigial mandibles but have well developed frontal glands are known as

(A) **Nasutes** (B) Pseudergates

(C) False workers (D) All of the above

169. Pseudergate/false workers in termites are found in which species

(A) *Odontotermes* (B) *Microtermes*

(C) ***Kalotermes*** (D) Mastotermes

170. The nest of termites is known as

(A) Mound (B) Termitaria

(C) Termitorium (D) **All of the above**

171. The king and queen are fed by the worker termites through

(A) **Trophollaxis** (B) Oral feeding

(C) Direct feeding (D) None of the above

172. For cellulose digetsion, the digestive tract of wood feeding termites contains

(A) Actinomycetes (B) **Trichonymphid Mastigophoran Protozoa**

(C) Bacteria (D) Fungi

173. The world's largest termite is

(A) *Kalotemes* sp. (B) ***Macrotermes bellicosus***

(C) *Odontotermes obesus* (D) *Mastotermes* sp.

174. The example of order Mallophaga is

(A) Bird lice (B) Chewing lice

(C) Biting lice (D) **All of the above**

175. The eggs of mallophaga is known as
(A) **Nit** (B) Egg mass
(C) Egg pod (D) Egg raft

176. The Mallophaga cling/stick to host with the help of
(A) Wings (B) **Mandibles**
(C) Antennae (D) Legs

177. The Anoplura/Siphunculata cling/stick to host with the help of
(A) Wings (B) Mandibles
(C) Antennae (D) **Legs**

178. The fore legs ofAnoplura/Siphunculata are modified as
(A) Raptorial legs (B) **Scansorial legs**
(C) Sticking legs (D) Suctorial legs

179. The blister or rashes caused by blister beetle is due to
(A) Meloidin (B) **Cantharidin**
(C) Lactic acid (D) None of the above

180. The scientific name of head louse is
(A) ***Pediculus humanus captis*** (B) *Pediculus humanus corporis*
(C) *Pthirus pubis* (D) Suctorial legs

181. The scientific name of Body louse is
(A) *Pediculus humanus captis* (B) ***Pediculus humanus corporis***
(C) *Pthirus pubis* (D) Suctorial legs

182. The scientific name of Crab louse is
(A) *Pediculus humanus captis* (B) *Pediculus humanus corporis*
(C) ***Pthirus pubis*** (D) Suctorial legs

183. The diseases transmitted by Body louse *Pediculus humanus corporis*is
(A) Epidemic typhus (B) Relapsing fever
(C) Trench fever (D) **All of the above**

184. The type of mouthparts in Hemiptera is
(A) Sponging type (B) Siphoning type
(C) Chewing and lapping type (D) **Piercing and sucking type**

185. In Hemiptera, the mouthparts are atrophied in
(A) Female coccids (B) **Male coccids**
(C) Male mealybug (D) Female melaybug

186. The presence of distinct Scutellum is characteristic of

(A) Hemiptera (B) **Homoptera**

(C) Heteroptera (D) None of the above

187. Filter chamber is present in

(A) Hemiptera (B) **Homoptera**

(C) Heteroptera (D) None of the above

188. In Hemiptera, the cerci is

(A) Present (B) **Absent**

(C) Present and single segmented (D) Present and multi segmented

189. In Hemiptera, Ovipositor is absent in

(A) Aphids (B) Coccids

(C) **Both A and B** (D) None of the above

190. In Cicadidae, the abdominal segment modified for sound production are

(A) First abdominal segment (B) **First 2 abdominal segment**

(C) First 3 abdominal segment (D) None of the above

191. In Hemiptera, the pupal stage is present in

(A) Males of Coccids (B) Whiteflies

(C) **Both A and B** (D) None of the above

192. The life cycle of periodical cicada is

(A) 50-75 days (B) 250-500 days

(C) 2-5 years (D) **13-17 years**

193. The spittle bugs/cuckoo spit/frog hopper belong to family

(A) Delphacidae (B) **Cercopidae**

(C) Membracidae (D) Lophopidae

194. The spittle froth in spittle bugs is formed by glands present on

(A) 4th and 5th abdominal segment

(B) 5th and 6th abdominal segment

(C) 6th and 7th abdominal segment

(D) **7th and 8th abdominal segment**

195. The wings are sloping over the abdomen in tent like manner in

(A) **Homoptera** (B) Heteroptera

(C) Cryptocerata (D) Gymnocerata

196. The wings overlapping (Flat) over the abdomen forming an "X" in

(A) Homoptera (B) **Heteroptera**

(C) Auchenorryncha (D) Sternorrhyncha

197. Assassin bug or Kissing bugs belong to family

(A) **Reduviidae** (B) Cimicidae

(C) Pentatomidae (D) Tingidae

198. Vector of *Trypanosoma cruzi* causing Trypanosomiasis is

(A) *Tritoma* sp. (B) *Rhodinus prolixus*

(C) Both A and B (D) None of the above

199. *Playmeris laevicollis* predator of coconut beetle (*Platymeris laevicollis*) belong to family

(A) **Reduviidae** (B) Cimicidae

(C) Pentatomidae (D) Tingidae

200. Bed bug belong to

(A) Phasmatodae (B) Anoplura

(C) **Hemiptera** (D) Mallophaga

201. Bed bug belong to family

(A) Reduviidae (B) **Cimicidae**

(C) Pentatomidae (D) Tingidae

202. The vector of Kala Azar is

(A) *Cimex lectularius* (B) *Cimex hemipterus*

(C) **Both A and B** (D) None of the above

203. Lace wings bug belong to family

(A) Reduviidae (B) Cimicidae

(C) Pentatomidae (D) **Tingidae**

204. The scientific name of Brinjal lace wing bug is

(A) *Stephanitis typicus* (B) ***Urentis hystericellus***

(C) *Tingis typicus* (D) None of the above

205. The scientific name of Banana lace wing bug is

(A) ***Stephanitis typicus*** (B) *Urentis hystericellus*

(C) *Tingis typicus* (D) None of the above

206. Cotton stainers/Red cotton bug belong to family

(A) Reduviidae (B) **Pyrrhocoreidae**
(C) Pentatomidae (D) Tingidae

207. Feeding on the boll by cotton stainers make the boll vulnerable for contamination by the fungus

(A) ***Nematospora lini*** (B) *Phytophthora lini*
(C) Both A and B (D) None of the above

208. The scientific name of red cotton bug is

(A) *Peregrinus maidis* (B) *Leptocorisa acuta*
(C) ***Dysdercus cingulatus*** (D) None of the above

209. Paddy gundhi bug belong to family

(A) Reduviidae (B) **Coreidae**
(C) Pentatomidae (D) Tingidae

210. Odoriferous glands of Paddy gundhi bug is present of the

(A) 2nd abdominal segment (B) 3rd abdominal segment
(C) 4th abdominal segment (D) **5th abdominal segment**

211. Odoriferous glands on the 5th abdominal segment produce nauseous odour due to

(A) Hex-*2-enal* (B) ***but-2-enal***
(C) *meth-2-enal* (D) *Ethane-2-enal*

212. Stink bug/ Shield bug belongs to family

(A) Reduviidae (B) Cimicidae
(C) **Pentatomidae** (D) Tingidae

213. Predatory bug, *Andrallus spinidens* of *Spodoptera litutra* belongs to family

(A) Reduviidae (B) Cimicidae
(C) **Pentatomidae** (D) Tingidae

214. Mustard Painted bug belong to family

(A) Reduviidae (B) Cimicidae
(C) **Pentatomidae** (D) Tingidae

215. The largest family of order Hemiptera is

(A) Reduviidae (B) Cimicidae
(C) Pentatomidae (D) **Cicadellidae/Jassidae**

216. Mango hopper belong to family
   (A) Reduviidae (B) Cimicidae
   (C) Pentatomidae (D) **Cicadellidae/Jassidae**

217. Vector of little leaf disease of brinjal is
   (A) ***Cestius pyhcitus*** (B) *Orosiusalbicinctus*
   (C) *Peregrinus maidis* (D) *Idioscopus niveopsparsus*

218. Rice green leaf hopper, *Nephotettix viriscens* is vector of
   (A) Yellow dwarf virus (B) Tungro virus
   (C) Dwarf virus (D) **All of the above**

219. Delphacids are characterised by
   (A) **Presence of Calcar at the apex of hind tibia**
   (B) Body raised up and the head almost touching the surface at rest
   (C) Vasiform orifice on the dorsal surface of last abdominal segment
   (D) Habit of running sideways

220. Cicadellids or Jassids are characterised by
   (A) Presence of Calcar at the apex of hind tibia
   (B) Body raised up and the head almost touching the surface at rest
   (C) Vasiform orifice on the dorsal surface of last abdominal segment
   (D) **Habit of running sideways**

221. Pysllids are characterised by
   (A) Presence of Calcar at the apex of hind tibia
   (B) **Body raised up and the head almost touching the surface at rest**
   (C) Vasiform orifice on the dorsal surface of last abdominal segment
   (D) Habit of running sideways

222. Aleyrodids are characterised by
   (A) Presence of Calcar at the apex of hind tibia
   (B) Body raised up and the head almost touching the surface at rest
   (C) **Vasiform orifice on the dorsal surface of last abdominal segment**
   (D) Habit of running sideways

223. Rice green leaf hopper, *Nephotettix viriscens* is vector of
   (A) Yellow dwarf virus (B) Tungro virus
   (C) Dwarf virus (D) **All of the above**

224. Hopper burn symptoms are typical of

(A) ***Nilaparvata lugens*** (B) *Nephotettix viriscens*

(C) *Sogatella furcifera* (D) *Recilia dorsalis*

225. Vector of Rice Yellow disease

(A) *Nilaparvata lugens* (B) *Nephotettix viriscens*

(C) ***Sogatella furcifera*** (D) *Recilia dorsalis*

226. Vector of Stripe disease of Jowar and Maize

(A) *Nilaparvata lugens* (B) *Nephotettix viriscens*

(C) *Sogatella furcifera* (D) ***Peregrinus maidis***

227. Vasiform Orifice is typical of

(A) Reduviidae (B) **Aleyrodidae**

(C) Pentatomidae (D) Cicadellidae

228. The scientific name of Sugarcane whitefly is

(A) *Aleurolobus barodensis* (B) *Diaphorina citri*

(C) *Bemesia tabaci* (D) ***Peregrinus maidis***

229. The scientific name of Citru psylla is

(A) *Aleurolobus barodensis* (B) ***Diaphorina citri***

(C) *Bemesia tabaci* (D) *Peregrinus maidis*

230. The pupal stage is present in

(A) Reduviidae (B) **Aleyrodidae**

(C) Pentatomidae (D) Cicadellidae

231. The Family Aphididae includes

(A) Aphids (B) Green lice

(C) Plant lice (D) **All of the above**

232. Cornicles (Honey tubes) are typically present in

(A) **Aphididae** (B) Aleyrodidae

(C) Pentatomidae (D) Cicadellidae

233. The life cycle of aphids pass through

(A) 5 stages (B) 6 stages

(C) 6 stages (D) **7 stages**

234. The different stages in life cycle of aphids are
(A) **Egg-Fundatrices-Fundatrigeniae-Migrants-Aliencolae-Sexuparae-Sexuales**
(B) Egg- Fundatrigeniae-Migrants - Fundatrices -Aliencolae-Sexuparae-Sexuales
(C) Egg- Migrants - Fundatrices - Fundatrigeniae- -Aliencolae-Sexuparae Sexuales
(D) Egg - Nymphal-Adult

235. Vector of Grassy shoot and mosaic virus in sugarcane is
(A) *Aphis craccivora* (B) ***Rhophalosiphum maidis***
(C) *Schizaphis graminum* (D) *Pentalonia nigronervosa*

236. Vector of Chirke virus in cardamom is
(A) *Aphis craccivora* (B) ***Rhophalosiphum maidis***
(C) *Schizaphis graminum* (D) *Pentalonia nigronervosa*

237. Vector of Bunchy top of bananais
(A) *Aphis craccivora* (B) *Rhophalosiphum maidis*
(C) *Schizaphis graminum* (D) ***Pentalonia nigronervosa***

238. The Banana aphid, *Rhophalosiphum maidis* is vector of
(A) Cardamom Mosaic virus (B) Bunchy top of banana
(C) Katte virus Cardamom (D) **All of the above**

239. The vector of Leaf roll and 'Y' viruses in potato is
(A) *Aphis craccivora* (B) ***Aphis gossypii***
(C) *Schizaphis graminum* (D) *Pentalonia nigronervosa*

240. The vector of common broad bean mosaic virus is
(A) ***Aphis craccivora*** (B) *Aphis gossypii*
(C) *Schizaphis graminum* (D) *Pentalonia nigronervosa*

241. Winter (roots) and summer (upper plant parts) migration is characteristic of
(A) *Aphis craccivora* (B) *Aphis gossypii*
(C) ***Eriosoma lanigerum*** (D) *Pentalonia nigronervosa*

242. The scientific name of Wooly apple aphid is
(A) *Aphis craccivora* (B) *Aphis gossypii*
(C) ***Eriosoma lanigerum*** (D) *Pentalonia nigronervosa*

243. Mango mealy bug belong to family

(A) Coccidae (B) Aleyrodidae

(C) **Margarodidae** (D) Cicadellidae

244. Apple cottony cushion scale belong to family

(A) Coccidae (B) Aleyrodidae

(C) **Margarodidae** (D) Cicadellidae

245. Apple cottony cushion scale, *Icerya purchasi* was first introduced in India in the year 1928 from

(A) Switzerland (B) **USA**

(C) Italy (D) England

246. The Cochineal bug belong to family

(A) **Dactylopidae** (B) Aleyrodidae

(C) Pseudococcidae (D) Cicadellidae

247. The Cochinael bug control

(A) Water hyacinth (B) **Prickly pear weed**

(C) Congress grass (D) Fox tail weed

248. Coffee green scale, *Coccus viridis* belong to family

(A) Aphididae (B) Aleyrodidae

(C) **Coccidae** (D) Pseudococcidae

249. Armoured scales belong to family

(A) Margarodidae (B) **Diaspididae**

(C) Coccidae (D) Pseudococcidae

250. San Jose Scale belong to family

(A) Margarodidae (B) **Diaspididae**

(C) Coccidae (D) Pseudococcidae

251. The scientific name of San Jose Scale is

(A) *Icerya purchasi* (B) ***Quadraspidiotus perniciosus***

(C) *Coccus viridis* (D) *Ceroplastes* sp.

252. The scientific name of Sugarcane leaf hopper is

(A) *Icerya purchasi* (B) *Quadraspidiotus perniciosus*

(C) ***Pyrilla perpusilla*** (D) *Ceroplastes* sp.

253. The Sugarcane leaf hopper belong to family

(A) Margarodidae (B) Diaspididae

(C) Coccidae (D) **Lophopidae**

254. The order Thysanoptera is characterised by

(A) Asymmetrical mouth parts (B) Fringed wings

(C) Presence of prepupal and pupal stage (D) **All of the above**

255. The mouth parts in thrips are

(A) **Piercing and sucking type** (B) Degenerate type

(C) Chewing and lapping type (D) None of the above

256. The suborder of Thysanoptera having ovipositor is

(A) Tubulifera (B) **Terebrantia**

(C) Both A and B (D) None of the above

257. Thesuborder of Thysanoptera having ovipositor absent is

(A) Tubulifera (B) **Terebrantia**

(C) Both A and B (D) None of the above

258. The wing veination is present in which suborder of Thysanoptera

(A) Tubulifera (B) **Terebrantia**

(C) Both A and B (D) None of the above

259. The Wing veination is Absent in which suborder of Thysanoptera

(A) **Tubulifera** (B) Terebrantia

(C) Both A and B (D) None of the above

260. The Thrips are

(A) **Xylem feeder** (B) Phloem feeder

(C) Both A and B (D) None of the above

261. The Ant lions, Mantispidflies, owlfies belong to order

(A) Trichoptera (B) **Neuroptera**

(C) Anoplura (D) Mallophaga

262. The Aphid lions, Golden eyes, green lacewing belong to order

(A) Trichoptera (B) **Neuroptera**

(C) Anoplura (D) Mallophaga

263. The Suborder Megaloptera and Planipennia belong to order

(A) Trichoptera (B) **Neuroptera**

(C) Anoplura (D) Mallophaga

264. The green lace wings, Aphid lion belong to family
   (A) Thripidae (B) **Chrysopidae**
   (C) Sphingidae (D) Mantispidae

265. The Mandibulosuctorial type mouth parts are present in Larvae of
   (A) Mantipidflies (B) **Lace wings**
   (C) Antlions (D) Owlflies

266. The stalked eggs are characteristic of
   (A) Mantipidflies (B) **Lace wings**
   (C) Antlions (D) Owlflies

267. The Moth, butterfly and skipper belong to order
   (A) Trichoptera (B) Neuroptera
   (C) Anoplura (D) **Lepidoptera**

268. The suborder Monotrysia and Ditrysia to order
   (A) Trichoptera (B) Neuroptera
   (C) Anoplura (D) **Lepidoptera**

269. The majority of Lepidopteran insect belong to sub order
   (A) **Ditrysia** (B) Monotrysia
   (C) Both A and B (D) None of the above

270. The dominant insects in order Lepidoptera
   (A) Butterflies (B) **Moths**
   (C) Skippers (D) All the above

271. The Mandibles are absent in order Lepidoptera except
   (A) Nymphalidae (B) **Micropterygidae**
   (C) Hespiridae (D) Noctuidae

272. The Distal cell in for ewings is characteristic of order
   (A) Trichoptera (B) Neuroptera
   (C) Anoplura (D) **Lepidoptera**

273. The Wing coupling mechanism in Lepidoptera is
   (A) Retinaculum (B) Amplexiform
   (C) Frenulum (D) **All the above**

274. The Moth, butterfly and skipper belong to order
   (A) Trichoptera (B) Neuroptera
   (C) Anoplura (D) **Lepidoptera**

275. The Pupa of butterfly is

(A) Obtect (B) Exarate

(C) **Chrysalis** (D) None of the above

276. The Four footed butterflies belong to family

(A) Lycaenidae (B) **Nymphalidae**

(C) Noctuidae (D) Hespiridae

277. The Four footed butterflies, the vestigial legs are

(A) **Forelegs** (B) Middle legs

(C) Hindlegs (D) None of the above

278. The anar butterfly or pomegranate fruit borer belong to family

(A) **Lycaenidae** (B) Nymphalidae

(C) Papilionidae (D) Hespiridae

279. The swallow tails belong to family

(A) Lycaenidae (B) Nymphalidae

(C) **Papilionidae** (D) Hespiridae

280. The Citrus buttlerfly, *Papilio demoleus* belong to family

(A) Lycaenidae (B) Nymphalidae

(C) **Papilionidae** (D) Hespiridae

281. The presence of Retractile Osmetaria at prothoracic tergum is typical of

(A) **Citrus butterfly larva** (B) Citrus butterfly Adult

(C) Both A and B (D) None of the above

282. The hind wings having tail like projections are typical of

(A) Lycaenidae (B) Nymphalidae

(C) **Papilionidae** (D) Hespiridae

283. The scientific name of cabbage butterfly is

(A) ***Pieris brassiace*** (B) *Melanitis leda ismene*

(C) *Virachola isocrates* (D) *Papilio demoleus*

284. The Rice horned caterpillar, *Melanitis leda ismene* belong to family

(A) **Satyridae** (B) Pieridae

(C) Arctiidae (D) Hespiridae

285. The Tiger moths belong to family

(A) Satyridae (B) Pieridae

(C) **Arctiidae** (D) Hespiridae

286. The Wooly bears belong to family
(A) Satyridae (B) Pieridae
(C) **Arctiidae** (D) Hespiridae

287. The scientific name of red hairy caterpillar is
(A) ***Amsacta albistriga*** (B) *Amsacta lactinea*
(C) *Spilosoma obliqua* (D) *Utethesia pulchella*

288. Thes cientific name of Bihar hairy caterpillar is
(A) *Amsacta albistriga* (B) *Amsacta lactinea*
(C) ***Spilosoma obliqua*** (D) *Utethesia pulchella*

289. The scientific name of Sunhemp hairy caterpillar is
(A) *Amsacta albistriga* (B) *Amsacta lactinea*
(C) *Spilosoma obliqua* (D) ***Utethesia pulchella***

290. The slug caterpillar belong to family
(A) Satyridae (B) **Cochlididae**
(C) Arctiidae (D) Hespiridae

291. The Snout moths belong to family
(A) Satyridae (B) Cochlididae
(C) Arctiidae (D) Pyralidae

292. Thes cientific name of Sorghum stem borer is
(A) *Scirpophaga incertulas* (B) ***Chilo partellus***
(C) *Cnaphalocrocis medinalis* (D *Agrotis ipsilon*

293. The example of Pyralidae is
(A) *Helulla undalis* (B) *Leucinoides orbonalis*
(C) *Nephopteryx eugraphylla* (D) **All the above**

294. The scientific name of sapota leaf webber is
(A) *Helulla undalis* (B) *Orthaga perticella*
(C) ***Nephopteryx eugraphylla*** (D) *Euzophera perticella*

295. The scientific name of Brinjal shoot and fruit borer is
(A) *Helulla undalis* (B) *Orthaga perticella*
(C) ***Leucinoide orbonalis*** (D) *Euzophera perticella*

296. The scientific name of Cabbage borer is
(A) ***Helulla undalis*** (B) *Orthaga perticella*
(C) *Nephopteryx eugraphylla* (D) *Euzophera perticella*

297. The scientific name of mango leaf webber is

(A) *Helulla undalis* (B) ***Orthaga perticella***

(C) *Nephopteryx eugraphylla* (D) *Euzophera perticella*

298. The Garin moths, Tuber moths paddy moth belong to family

(A) **Gelechiidae** (B) Pieridae

(C) Arctiidae (D) Hespiridae

299. The Trapezoidal forewings are characteristics of

(A) **Gelechiidae** (B) Pieridae

(C) Arctiidae (D) Hespiridae

300. The Cotton pink bollworm, *Pectinophora gossypiella* belong to family

(A) Satyridae (B) Pieridae

(C) **Gelechiidae** (D) Hespiridae

301. The Potato tuber moth, *Phthormoea opeculella* belong to family

(A) Satyridae (B) Pieridae

(C) **Gelechiidae** (D) Hespiridae

302. The Groundnut leaf miner, *Aproaerema modicella* belong to family

(A) Satyridae (B) Pieridae

(C) **Gelechiidae** (D) Hespiridae

303. The Looper belong to family

(A) Satyridae (B) **Geometridae**

(C) Arctiidae (D) Hespiridae

304. The Loopers are also called as

(A) Inch worm (B) Measuring worm

(C) Earth measurer (D) **All the above**

305. The Tussock moths belong to family

(A) **Lymantridae** (B) Pieridae

(C) Noctuidae (D) Hespiridae

306. The largest family of order Lepidoptera is

(A) Satyridae (B) Pieridae

(C) **Noctuidae** (D) Hespiridae

307. The Semi loopers belong to family

(A) Satyridae (B) Pieridae

(C) **Noctuidae** (D) Hespiridae

308. The Cutworms belong to family

(A) Satyridae (B) Pieridae

(C) **Noctuidae** (D) Hespiridae

309. The Armyworms belong to family

(A) Satyridae (B) Pieridae

(C) **Noctuidae** (D) Hespiridae

310. The Fruit sucking moths belong to family

(A) Satyridae (B) Pieridae

(C) **Noctuidae** (D) Hespiridae

311. The Noctuids pupate in the

(A) **Soil** (B) Plants

(C) Leaves (D) Fruits

312. The Lepidoptera insect having adults as damaging stage

(A) **Fruit sucking moth** (B) Tussock moth

(C) Sphinx moth (D) Rice skipper

313. The *Scirpophaga incertulas* belong to family

(A) Satyridae (B) **Pyraustidae**

(C) Noctuidae (D) Hespiridae

314. The *Antherea pahia* is

(A) **Giant silk moth** (B) Grass borer

(C) Horn worm (D) Whites

315. The Tassar silk moth, *Antherea pahia* belong family

(A) Satyridae (B) Pieridae

(C) Noctuidae (D) **Saturnidae**

316. The Hawk moths, Sphinx moths, Horn moths belong to family

(A) Satyridae (B) Pieridae

(C) **Sphingidae** (D) Hespiridae

317. In Sphingids, conspicuous horn is present on dorsal surface of

(A) 5th Abdominal segment (B) 6th Abdominal segment

(C) 7th Abdominal segment (D) **8th Abdominal segment**

318. The scientific name of Death's Head Moth

(A) *Pelopidas mathias* (B) ***Acherontia styx***

(C) *Otheris materna* (D) *Scirpophaga incertulas*

319. The scientific name of Rice skipper is
(A) ***Pelopidas mathias*** (B) *Acherontia styx*
(C) *Otheris materna* (D) *Scirpophaga incertulas*

320. The Skippers belong to family
(A) Satyridae (B) Pieridae
(C) Sphingidae (D) **Hespiridae**

321. The Skippers are characterised by
(A) Antenna is apically prolonged beyond the club into a hook
(B) Wings held at different directions at rest
(C) Erratic and darting flight
(D) **All the above**

322. The scientific name of Death's Head Moth
(A) *Pelopidas mathias* (B) ***Acherontia styx***
(C) *Otheris materna* (D) *Scirpophaga incertulas*

323. The examples of order Diptera is
(A) Mosquitoes (B) Gnats
(C) Midges (D) **All the above**

324. The wings modified as Haltere in Diptera are
(A) Forewings (B) **Hind wings**
(C) Both Fore and hindwings (D) All the above

325. The Larvae of Dipetra are called as
(A) **Maggots** (B) Loopers
(C) Eruciform (D) Grubs

326. The Pupa of Diptera is
(A) Exarate (B) Obtect
(C) **Coarctate** (D) Decticuous

327. The Sub order of Diptera is
(A) Nematocera (B) Brachycera
(C) Cyclorrhapha (D) **All the above**

328. The Eucephalus head is characteristic of suborder
(A) **Nematocera** (B) Brachycera
(C) Cyclorrhapha (D) All the above

329. The Hemicephalus head is characteristic of suborder

(A) Nematocera (B) **Brachycera**

(C) Cyclorrhapha (D) All the above

330. The Acephalus head is characteristic of suborder

(A) Nematocera (B) Brachycera

(C) **Cyclorrhapha** (D) All the above

331. The eversible bladder Ptilinum or Frontal sac present in the head is characteristic of

(A) Nematocera (B) Brachycera

(C) **Cyclorrhapha** (D) All the above

332. The larvae of mosquitoes are known as

(A) **Wriggler** (B) Tumbler

(C) Maggot (D) None of the above

333. The Pupae of mosquitoes are known as

(A) Wriggler (B) **Tumbler**

(C) Maggot (D) None of the above

334. The gall midges belong to family

(A) **Ceccidomyidae** (B) Agromyzidae

(C) Tabanidae (D) Syrphidae

335. A dark sclerotised area is present midventrally on the pro-thorax in the larva called 'Chestbone'is characteristic of

(A) **Ceccidomyidae** (B) Agromyzidae

(C) Tabanidae (D) Syrphidae

336. The Robber flies belong to family

(A) **Asilidae** (B) Agromyzidae

(C) Tabanidae (D) Syrphidae

337. The Horse flies belong to family

(A) Ceccidomyidae (B) Agromyzidae

(C) **Tabanidae** (D) Syrphidae

338. The Anthrax is vectored by

(A) Robber flies (B) **Horse flies**

(C) Syrphid flies (D) Gnats

339. The Hover flies belong to family
(A) Ceccidomyidae (B) Agromyzidae
(C) Tabanidae (D) **Syrphidae**

340. The Fruit flies belong to family
(A) **Tephritidae** (B) Agromyzidae
(C) Tabanidae (D) Syrphidae

341. The Dipteran mostly used in study of animal genetics
(A) *Dacus cucurbitae* (B) ***Drosophila melanogatser***
(C) Tachinid flies (D) *Musca domestica*

342. The Vinegar gnats, Pomace flies belong to family
(A) **Drosophildae** (B) Agromyzidae
(C) Tabanidae (D) Syrphidae

343. The maggots of house fly is
(A) **Scavanger** (B) Carnivorous
(C) herbivorous (D) Fluid feeder

344. The Dog flies belong to family
(A) Tephritidae (B) **Hippoboscidae**
(C) Tabanidae (D) Syrphidae

345. The Leaf miners belong to family
(A) Tephritidae (B) **Agromyzidae**
(C) Tabanidae (D) Syrphidae

346. The Scientific name of Pea stem fly is
(A) *Melanogromyza obtusa* (B) *Liriomyza trifolii*
(C) ***Ophiomyia phaseoli*** (D) None of the above

347. The Scientific name of Serpentine leaf miner is
(A) *Melanogromyza obtusa* (B) ***Liriomyza trifolii***
(C) ***phiomyia phaseoli*** (D) None of the above

348. The Shoot flies belong to family
(A) Tephritidae (B) Agromyzidae
(C) **Muscidae** (D) Syrphidae

349. The typical characteristic of Hymenoptera is
(A) **Hamulate wing coupling** (B) Presence of ovipositor
(C) Complete metamorphosis (D) **All the above**

350. The examples of Hymenoptera is

(A) wasp (B) Ants

(C) Bees (D) **All the above**

351. The Chewing and lapping types mouthparts are present in

(A) **Hymenoptera** (B) Diptera

(C) Thysanoptera (D) Coleoptera

352. The Propodeum and Gatser are present in

(A) **Hymenoptera** (B) Diptera

(C) thysanoptera (D) Coleoptera

353. The phytophagous suborder of hymenoptera is

(A) Apocrita (B) **Symphyta**

(C) None of the above (D) All the above

354. The sawflies belong to family

(A) **Tenthridinidae** (B) Icheneumonidae

(C) Braconidae (D) Chalcididae

355. The very long ovipositor often longer than body and permanently extruded is characteristic of

(A) Tenthridinidae (B) **Icheneumonidae**

(C) Braconidae (D) Chalcididae

356. *Isotima javensis*, Larval parasitoid of sugarcane top shoot borer belong to family

(A) Tenthridinidae (B) **Icheneumonidae**

(C) Braconidae (D) Chalcididae

357. The parasitoids which pupate in the silken cocoon on the outside the host belong to family

(A) Tenthridinidae (B) Icheneumonidae

(C) **Braconidae** (D) Chalcididae

358. *Apanteles* (*Cotesia*) *glomeratus,* arval parasitoid of cabbage caterpillar belong to family

(A) Tenthridinidae (B) Icheneumonidae

(C) **Braconidae** (D) Chalcididae

359. The number of prolegs in sawfly is

(A) 5 pairs (B) 6 pairs

(C) 7 pairs (D) **8 pairs**

360. The egg parasitic family is

(A) **Trichogrammitidae** (B) Chalcididae

(C) Bethylidae (D) Eulophidae

361. The parasitiod extensively used worldwide for augmentative biological control programmes is

(A) ***Trichogramma*** (B) *Cotesia*

(C) *Bracon* (D) *Isotima*

362. The Yellow Jackets, Hornets belong to family

(A) Tenthridinidae (B) Sphecidae

(C) Braconidae (D) **Vespidae**

363. TheAnts belong to family

(A) **Formicidae** (B) Xylocopidae

(C) Apidae (D) Megachilidae

364. The honeybee belong to family

(A) Formicidae (B) Xylocopidae

(C) **Apidae** (D) Megachilidae

365. The Leaf cutter belong to family

(A) Formicidae (B) Xylocopidae

(C) Apidae (D) **Megachilidae**

366. The carpenter bee belong to family

(A) Formicidae (B) **Xylocopidae**

(C) Apidae (D) Megachilidae

367. The Adephaga and Polyphaga belong to order

(A) Trichoptera (B) Lepidoptera

(C) Dermaptera (D) **Coleoptera**

368. The Elytra is characteristic of order

(A) Trichoptera (B) Lepidoptera

(C) Dermaptera (D) **Coleoptera**

369. The Tiger beetle belong to family

(A) **Cicindelidae** (B) Carabidae

(C) Curculinoidae (D) Cerambycidae

370. The Ground beetle belong to family

(A) Brichidae (B) **Carabidae**

(C) Curculinoidae (D) Cerambycidae

371. The true water beetle belong to family
(A) **Dytiscidae** (B) Hydrophildae
(C) Curculinoidae (D) Cerambycidae

372. The Whirligig beetle belong to family
(A) **Gyrinidae** (B) Coccinellidae
(C) Curculinoidae (D) Cerambycidae

373. The Fireflies or Glow worms belong to family
(A) Gyrinidae (B) **Lampyridae**
(C) Curculinoidae (D) Cerambycidae

374. The Cigarette beetle, *Lasioderma serricornae* belong to family
(A) Bruchidae (B) **Anobiidae**
(C) Curculinoidae (D) Cerambycidae

375. The Lesser grain borer, *Rhizopertha dominica* belong to family
(A) Bruchidae (B) Anobiidae
(C) Curculinoidae (D) **Bostrychidae**

376. The Pulse beetle, *Callosobruchus chinensis* belong to family
(A) **Bruchidae** (B) Coccinellidae
(C) Curculinoidae (D) Tenebrionidae

377. The red flour beetle, *Tribolium castaneum* belong to family
(A) Bruchidae (B) Coccinellidae
(C) Curculinoidae (D) **Tenebrionidae**

378. The Hadda beetle belong to family
(A) Brichidae (B) **Coccinellidae**
(C) Curculinoidae (D) Cerambycidae

379. The Lady bird beetle belong to family
(A) Brichidae (B) **Coccinellidae**
(C) Curculinoidae (D) Cerambycidae

380. *Rodolia cardinalis,* predator of cottony cushion scale belong to family
(A) Brichidae (B) **Coccinellidae**
(C) Curculinoidae (D) Cerambycidae

381. The characteristic of Bruchidae
(A) Serrate antennae (B) Elytra short not cover tip of abdomen
(C) Hypermetamorphosis (D) **All the above**

382. TheWeevils belong to family

(A) Brichidae (B) Coccinellidae

(C) **Curculinoidae** (D) Cerambycidae

383. The characteristic of Curculionids is

(A) Rostrum/snout (B) Geniculate antennae

(C) Grubs apodus (D) **All the above**

384. The scientific name of sweet potato weevil is

(A) *Cosmopolites sordidus* (B) ***Cylas formicarius***

(C) *Rhynchophorus ferrugines* (D) *Sitophilus oryzae*

385. Thescientific name of banana rhizome weevil is

(A) ***Cosmopolites sordidus*** (B) *Cylas formicarius*

(C) *Rhynchophorus ferrugines* (D) *Sitophilus oryzae*

386. The scientific name of Rice weevil is

(A) *Cosmopolites sordidus* (B) *Cylas formicarius*

(C) *Rhynchophorus ferrugines* (D) ***Sitophilus oryzae***

387. The scientific name of coconut red palm weevil is

(A) *Cosmopolites sordidus* (B) *Cylas formicarius*

(C) ***Rhynchophorus ferrugines*** (D) *Sitophilus oryzae*

388. Thecharacteristic of Scarabaeidae is

(A) Head often with toothed/bifurcated frontal horn

(B) Elytra usually not cover abdomen completely

(C) Grubs Scarabaeiform

(D) **All the above**

389. TheScarabaeidae are commonly known as

(A) June beetle (B) Root grubs

(C) White grubs (D) **All the above**

390. The scientific name of coconut rhinoceros beetle is

(A) *Anomala varians* (B) *Holotrichia insularis*

(C) *Holotrichia serrata* (D) ***Oryctus rhinoceros***

391. The Flea beetle, Leaf beetle or Tortoise beetle belong to family

(A) Cerambycidae (B) Anobiidae

(C) **Chyrsomelidae** (D) Scarabaeidae

392. The Red pumpkin beetle beetle belong to family

(A) Cerambycidae (B) Anobiidae

(C) **Chyrsomelidae** (D) Scarabaeidae

393. The Rice Hispa belong to family

(A) Cerambycidae (B) Anobiidae

(C) **Chyrsomelidae** (D) Scarabaeidae

394. The Cerambycidae is characterised by

(A) Very long filiform antennae at least 2/3rd of body length

(B) Prothorax usually spined or tuberculate

(C) Tarsal formula 5-5-5

(D) **All the above**

395. The scientific name of Mango stem borer is

(A) *Cosmopolites sordidus* (B) ***Batocera rufomaculata***

(C) *Plocaederus ferrugieus* (D) *Sthenias grisator*

396. The scientific name of grapevine stem girdler is

(A) *Cosmopolites sordidus* (B) *Batocera rufomaculata*

(C) *Plocaederus ferrugieus* (D) ***Sthenias grisator***

397. The scientific name of Cashew Tree borer is

(A) *Cosmopolites sordidus* (B) *Batocera rufomaculata*

(C) ***Plocaederus ferrugieus*** (D) *Sthenias grisator*

398. The Coconut Rhinoceros beetle belong to family

(A) Cerambycidae (B) Anobiidae

(C) Chyrsomelidae (D) **Dynastidae**

399. The Click beetle, wire worms belong to family

(A) **Elatteridae** (B) Anobiidae

(C) Chyrsomelidae (D) Dynastidae

400. The scientific name of red pumpkin beetle is

(A) *Holocera pulvicera* (B) *Mylabris pustulata*

(C) *Galerucella birmanica* (D) ***Aulacophora foveicollis***

401. The example of Melolonthidae family

(A) June beetle (B) Chaffer beetle

(C) hite grubs (D) **All of the above**

402. The blister beetle belong to family

(A) Cerambycidae (B) Anobiidae

(C) **Meloidae** (D) Dynastidae

403. The scientific name of banded blister beetle is

(A) *Holocera pulvicera* (B) ***Mylabris pustulata***

(C) *Galerucella birmanica* (D) *Aulacophora foveicollis*

Unit-4

Insect Ecology

Q 1. Match the following

1. The Father of Ecology is

(A) Tansley (B) **Odum**

(C) Cowley (D) R. F. Chapman

2. The term Ecology is derived from

(A) Italic (B) Latin

(C) **Greek** (D) French

3. The term Ecology was coined by

(A) Reiter (B) Odum

(C) Cowley (D) **Haeckel**

4. The term Ecosysetm was coined by

(A) **Tansley** (B) Odum

(C) Cowley (D) Haeckel

5. The author of "*Insect Ecology*" is

(A) **P.W. Price** (B) T.R.E. Southwood and P.A.Henderson

(C) C. J. Krebs (D) M.R.Speight, M.D.Hunta and A.D. Watt

6. The author of "*Ecology of Insects : Concepts and Application*" is

(A) P.W. Price (B) T.R.E. Southwood and P.A. Henderson

(C) C. J. Krebs (D) **M.R.Speight, M.D.Hunta and A.D. Watt**

7. The author of " *Ecological Methods*" is

(A) P.W. Price (B) **T.R.E. Southwood and P.A. H enderson**

(C) C. J. Krebs (D) M.R.Speight, M.D.Hunta and A.D. Watt

8. The study of the structure and functions of nature

(A) Autoecology (B) Synecology
(C) Both A and B (D) **Ecology**

9. The study of individual organisms or an individual species in relation to the environment is

(A) **Autoecology** (B) Synecology
(C) Both A and B (D) Ecology

10. The study of group of organisms in relation to the environment is

(A) Autoecology (B) **Synecology**
(C) Both A and B (D) Ecology

11. Groups of individuals of any kind of organisms of the same species occupying a particular area at a specific time

(A) Ecosystem (B) **Population**
(C) Biosphere (D) Community

12. Groups of species of organisms found at given place at particular point of time

(A) Ecosystem (B) Population
(C) Biosphere (D) **Community**

13. An assemblage of interacting plant and animal communities and non living environmental components

(A) **Ecosystem** (B) Population
(C) Biosphere (D) ommunity

14. The process of repeated eating and being eaten is known as

(A) Food web (B) **Food Chain**
(C) Niche (D) All the above

15. A network of inter related food chains is known as

(A) **Food web** (B) Food Chain
(C) Niche (D) All the above

16. The innate ability of the population to reproduce and survive is termed as

(A) **Biotic potential** (B) Potential Natality
(C) Both A and B (D) Potential Fecundity

17. The capacity of environment to support and sustain a level of population under an ideal condition is termed as

(A) Biotic potential (B) Potential Natality
(C) **Carrying capacity** (D) Potential Fecundity

18. The reproductive rate of the individuals in an optimal environment is termed as

(A) Biotic potential (B) **Potential Natality**

(C) Both A and B (D) Potential Fecundity

19. The rate at which new individuals are added to the population by reproduction

(A) Biotic potential (B) **Natality**

(C) Mortality (D) Fecundity

20. The rate at which individuals are lost by death

(A) Biotic potential (B) Natality

(C) **Mortality** (D) Fecundity

21. The rate at which individuals immigrate into and emigrate out of the population

(A) **Dispersal** (B) Natality

(C) Mortality (D) Fecundity

22. The way in which individuals are distributed in space

(A) Dispersal (B) **Dispersion**

(C) Immigrate (D) Emigrate

23. The way in which individuals are scattered in groups

(A) Uniform distribution (B) **Clumped distribution**

(C) Random distribution (D) Emigrate

24. The species characterized by low reproductive rate and high survival rate are

(A) **k-strategist** (B) r-strategist

(C) Both A and B (D) None the above

25. The species characterized by high reproductive rate and low survival rate are

(A) k- strategist (B) **r-strategist**

(C) Both A and B (D) None the above

26. Which one the following is r-strategist

(A) **Aphids** (B) *Helicoverpa*

(C) Coddling moth (D) All of the above

27. Which one the following is r-strategist

(A) **Aphids** (B) *Helicoverpa*

(C) Coddling moth (D) All of the above

28. The growth form of r-strategists follow

(A) **J-shape growth form** (B) S-shape growth form

(C) L-shape growth form (D) All of the above

29. The insects having cannabalism properties

(A) *Helicoverpa* (B) Chrysopids
(C) Tribolium confusum (D) All of the above

30. The overcrowding in honeybees lead to

(A) **Swarming** (B) Absconding
(C) Migration (D) All of the above

31. The overcrowding in rice weevil lead to

(A) Alate forms (B) **Reduction in fecundity**
(C) Migration (D) All of the above

32. The overcrowding in Aphids lead to

(A) **Alate forms for migration** (B) Reduction in fecundity
(C) Dispersal (D) All of the above

33. The elimination of one species by another as a result of interspecific competition has come to be known as

(A) Competitive exclusion principle (B) Gause's Principle
(C) **Both A and B** (D) None of the above

34. *Tribolium castaneum* and *T. confusum* when reared in the same jar at high temperature and RH

(A) ***T. castaneum* will eliminate *T. confusum***
(B) *T. confusum* will eliminate *T. castaneum*
(C) Both *T. confusum* and *T. castaneum* will survive together
(D) All of the above

35. *Tribolium castaneum* and *T. confusum* when reared in the same jar at Low temperature and RH

(A) *T. castaneum* will eliminate *T. confusum*
(B) ***T. confusum* will eliminate *T. castaneum***
(C) Both *T. confusum* and *T. castaneum* will survive together
(D) All of the above

36. The total heat required for completion of physiological processes in life history

(A) **Thermal Constant** (B) Thermal heat constant
(C) Metabolic heat constant (D) All of the above

37. The Thermal constant for a particular insect is

(A) Variable (B) **Fixed**
(C) Both A and B (D) None of the above

38. The According to Vat Hoff's Law, if the temperature is increased by 10ºC, the rate of development is

(A) **Doubled** (B) Three times
(C) Quadruple (D) Six times

39. In general, the insects in cooler areas are

(A) Dark coloured (B) Light coloured
(C) Both A and B (D) None of the above

40. In desert insects the light colour helps

(A) Reflect sunlight (B) Disperse sunlight
(C) Refract sunlight (D) **All of the above**

41. Oriented locomotory movement towards or away the from the source of light is termed as

(A) Phototropism (B) **Phototaxis**
(C) Photoperiodism (D) All of the above

42. Light directed growth mechanism in sessile insects is termed as

(A) **Phototropism** (B) Phototaxis
(C) Photoperiodism (D) All of the above

43. The response of insects to environmental rythms of light and darkness is termed as

(A) Phototropism (B) Phototaxis
(C) **Photoperiodism** (D) All of the above

44. The Scotophase is the

(A) **Period of darkness** (B) Period of light/illumination
(C) Both A and B (D) None of the above

45. The Photophase is the

(A) Period of darkness (B) **Period of light/illumination**
(C) Both A and B (D) None of the above

46. The High Relative humidity predisposes

(A) BPH attack in rice (B) Aphid attack
(C) *Verticllium* attack on *coccus viridis* (D) **All of the above**

47. The insect active at night are

(A) **Nocturnal** (B) Crepuscular
(C) Diurnal (D) All of the above

48. The insect active at day are

(A) Nocturnal (B) Crepuscular
(C) **Diurnal** (D) All of the above

49. The insect active at during dawn and dusk are

(A) Nocturnal (B) **Crepuscular**
(C) Diurnal (D) All of the above

50. The rainfall is essential for the emergence of

(A) White grubs (B) Cutworms
(C) Red Hairy caterpillar (D) **All of the above**

51. The tabular statements showing the number of insects dying over a period of time and accounting for their deaths

(A) **LifeTable** (B) Growth table
(C) Growth stage table (D) All of the above

52. The physical and biological restraints that prevent a species from realizing its Biotic potential

(A) **Environmental resistance** (B) Biotic resistance
(C) Both A and B (D) None of the above

53. The competition between individuals of same species for resources is

(A) **Intraspecific competition** (B) Interspecific competition
(C) Both A and B (D) All of the above

54. The competition between individuals of different species for resources is

(A) Intraspecific competition (B) **Interspecific competition**
(C) Both A and B (D) All of the above

55. The Sympatric population shows

(A) Population whose geographic ranges overlap
(B) **Population showing similar adaptations**
(C) Population showing dissimilar adaptations
(D) Population at verge of extinction

56. The Facultative factor are

(A) Parasitoid (B) Food
(C) redators (D) **All the above**

57. The Density dependent factor are

(A) Parasitoid (B) Food
(C) Predators (D) **All the above**

58. The Catastrophic factor are
    (A) Temperature (B) Humidity
    (C) Rainfall (D) **All the above**

59. The Density independent factor are
    (A) Temperature (B) Humidity
    (C) Rainfall (D) **All the above**

60. The Insects are
    (A) Homeothermic (B) **Poikilothermic**
    (C) Socio-Thermic (D) None the above

61. The Homeothermic animals are also known as
    (A) Ectothermic (B) **Endothermic**
    (C) Socio-Thermic (D) None the above

62. The Poikilothermic animals are also known as
    (A) **Ectothermic** (B) Endothermic
    (C) Socio-Thermic (D) None the above

63. The Insects are
    (A) Homeothermic (B) **Poikilothermic**
    (C) Socio-Thermic (D) None the above

64. The Honeybees are
    (A) Homeothermic (B) Poikilothermic
    (C) **Socio-Thermic** (D) None the above

65. The animals which maintain a constant body temperature within certain narrow limits irrespective of the temperature variations in the external nvironment.
    (A) **Homeothermic** (B) Poikilothermic
    (C) Socio-Thermic (D) None the above

66. The animals which are able to maintain a body temperature in the colony /nest with respect of the temperature variations in the external environment.
    (A) Homeothermic (B) Poikilothermic
    (C) **Socio-Thermic** (D) None the above

67. The animals which are unable to maintain a constant body temperature with respect of the temperature variations in the external environment.
    (A) Homeothermic (B) **Poikilothermic**
    (C) Socio-Thermic (D) None the above

68. The temperature range for Zone of optimum or Normal development is .
    (A) 10-25°C (B) **10-35°C**
    (C) 10-40°C (D) None the above

69. The Summer sleep or period of suspended activity in summer is
    (A) **Aestivation** (B) Diapause
    (C) Hibernation (D) Quiescence

70. The Winter sleep or period of suspended activity in winter is
    (A) Aestivation (B) Diapause
    (C) **Hibernation** (D) Quiescence

71. Which one of these is genetically determined
    (A) Aestivation (B) **Diapause**
    (C) Hibernation (D) Quiescence

72. The temporary period of suspended activity in due to sudden variation in temperature is
    (A) Aestivation (B) Diapause
    (C) Hibernation (D) **Quiescence**

73. The temperature range for Zone of optimum or Normal development is .
    (A) 10-25°C (B) **10-35°C**
    (C) 10-40°C (D) None the above

74. The cold hardiness of insects is due to
    (A) Glycerol (B) Sorbitol
    (C) Erythritol (D) **All of the above**

75. Biomagnification of pesticides is highest in
    (A) **Terminal** level (B) Low level
    (C) Middle level (D) All level

76. Prepupal diapause is present in
    (A) *Bombyx* (B) *Pieris brassiace*
    (C) ***Plodia interpunctella*** (D) *Euproctis*

77. Pupal diapause is present in
    (A) *Bombyx* (B) ***Pieris brassiace***
    (C) *Plodia interpunctella* (D) *Euproctis*

78. Larval diapause is present in
    (A) *Bombyx* (B) *Pieris brassiace*
    (C) *Plodia interpunctella* (D) ***Euproctis***

79. Embryonic diapause is observed in

(A) ***Bombyx*** (B) *Leptinotarsa*

(C) *Hyalophora* (D) *Glossina*

80. The insect undergo diapause in egg stage is

(A) *Scirpophaga incertulas* (B) *Leptinotarsa decimlineata*

(C) **Grasshopper** (D) *Lymantria dispar*

81. The insect undergo diapause in adult stage is

(A) *Scirpophaga incertulas* (B) ***Leptinotarsa decimlineata***

(C) *Grasshopper* (D) *Lymantria dispar*

82. The insect undergo diapause in larval stage is

(A) ***Scirpophaga incertulas*** (B) *Leptinotarsa decimlineata*

(C) *Grasshopper* (D) *Lymantria dispar*

83. The insect undergo diapause in Late embryonic stage is

(A) *Scirpophaga incertulas* (B) *Leptinotarsa decimlineata*

(C) *Grasshopper* (D) ***Lymantria dispar***

84. Imaginal diapause is present in

(A) *Culex* (B) *Anopheles*

(C) **Both A and B** (D) *Euproctis*

85. The type of soil preferred by insects

(A) Clay loam (B) **Loamy soil**

(C) Sandy soil (D) Clay soil

86. The tendency of insects to concentrate in number larger than found in normal distribution is termed as

(A) **Aggregation** (B) Concurrence

(C) Association (D) All of the above

87. The mass assemblage of insects at one time is termed as

(A) Aggregation (B) **Concurrence**

(C) Association (D) All of the above

88. The type of mechanism in which adverse effect of the host plant on the biology of the insect pest.

(A) Non preference (B) Antixenosis

(C) Antibiosis (D) **All of the above**

89. A relationship in which one partner is benefitted, the other remaining unbenefitted and unharmed is

(A) Cooperation (B) Mutualism

(C) **Commensalism** (D) Disoperation

90. The interaction where one species has a negative effect on the other, but there is no measurable reciprocal effect

(A) Antagonism (B) Mutualism

(C) Commensalism (D) **Amensalism**

91. The interaction where one species has a negative effect on the other species , or both species exert a negative effect

(A) **Antagonism** (B) Mutualism

(C) Commensalism (D) Amensalism

92. The exploitation of a common and limiting resource by two or more species is

(A) Cooperation (B) Mutualism

(C) Commensalism (D) **Competition**

93. One species enters the domicile of another habitually, with or without causing damage to the host species

(A) Cooperation (B) Mutualism

(C) **Inquilnlism** (D) Competition

94. The close association of two or more species living together, no matter which kind of influence one has on the other

(A) Cooperation (B) Mutualism

(C) **Symbiosis** (D) Competition

95. Both species benefit from an association

(A) Cooperation (B) **Mutualism**

(C) Symbiosis (D) Competition

96. Individuals of one species kill and eat individuals of another species

(A) Pathogen (B) **Predator**

(C) Parasitoid (D) Parasite

97. Which one of these are beneficial for controlling insect-pests

(A) **Primary parasitoid** (B) Secondary parasitoid

(C) Hyperparasite (D) All the above

98. The rate at which a population increases in size if there are no density dependent factors regulating the population is known as
   (A) Extrinsic rate of Increase (B) **Intrinsic rate of increase**
   (C) Both A and B (D) None of the above

99. The depopulation or movement of organisms out of the area is
   (A) **Emigration** (B) Dispersal
   (C) Immigration (D) Migration

100. The movement of organisms in to the area from outside is
   (A) Emigration (B) Dispersal
   (C) **Immigration** (D) Migration

101. The mass movement of large of entire population is
   (A) Emigration (B) Dispersal
   (C) Immigration (D) **Migration**

102. The measure of change in the reproductive capacity from generation ot generation is termed as
   (A) **Replacement Rate(R)** (B) Generation rate
   (C) Change rate (D) All of the above

103. The value of Replacement Rate (R) <1 indicate that
   (A) **The population is decreasing** (B) The population is increasing
   (C) The population is stable (D) All of the above

104. The value of Replacement Rate (R) >1 indicate that
   (A) The population is decreasing (B) **The population is increasing**
   (C) The population is stable (D) All of the above

105. The value of Replacement Rate (R) =1 indicate that
   (A) The population is decreasing (B) The population is increasing
   (C) **The population is stable** (D) All of the above

106. The Lentic aquatic ecosystem include
   (A) Pond (B) Lakes
   (C) Pools (D) **All of the above**

107. The Lotic aquatic ecosystem include
   (A) streams (B) Rivers
   (C) Canals (D) **All of the above**

108. The respiratory structures in aquatic insects include
(A) Tracheal gills (B) Respiratory siphon
(C) Plastron (D) **All of the above**

109. The term Ecological Niche was coined by
(A) Haeckel (B) Tansley
(C) Odum (D) **Grinnel**

110. The set of abiotic and biotic conditions under which a species or population is able to persist and maintain stable population is termed as
(A) **Niche** (B) Fundamental Niche
(C) Realized Niche (D) All of the above

111. The set of environmental conditions under which a species or population is able to persist and maintain stable population is termed as
(A) Ecological Niche (B) **Fundamental Niche**
(C) Realized Niche (D) All of the above

112. The set of set of environmental plus ecological conditions under which a species or population is able to persist and maintain stable population is termed as
(A) Niche (B) Fundamental Niche
(C) **Realized Niche** (D) Ecological Niche

113. The functional role of the species in its community and the position of the species of the species along all environment gradients termed as
(A) Niche **(B) Fundamental Niche**
(C) Realized Niche (D) **Ecological Niche**

114. The Life table concept was first given by
(A) Morris and Miller (B) Raymond Pearl
(C) R. J. Tansley (D) None of the above

115. The First Life table for natural insect population was given by
(A) **Morris and Miller** (B) Raymond Pearl
(C) R. J. Tansley (D) None of the above

116. The Life table for natural insect population was studied in
(A) *Helicoverpa armigera* (B) ***Choristoneura fumiferana***
(C) *Plodia Interpunctella* (D) None of the above

117. The Life table studying organism mortality and reproduction as function of stage is
   - (A) Age specific Life table
   - (B) **Stage specific Life table**
   - (C) Both A and B
   - (D) None of the above

118. The Life table studying organism mortality and reproduction as function of age is
   - (A) **Age specific Life table**
   - (B) Stage specific Life table
   - (C) Both A and B
   - (D) None of the above

119. The Age specific Life table is also known as
   - (A) Vertical Life table
   - (B) **Horizontal Life table**
   - (C) Both A and B
   - (D) None of the above

120. The Stage specific Life table is also known as
   - (A) **Vertical Life table**
   - (B) Horizontal Life table
   - (C) Both A and B
   - (D) None of the above

121. In life tables, type I curve shows
   - (A) Low death rate of young
   - (B) High mortality in old
   - (C) **Both A and B**
   - (D) Uniform mortality of young and old

122. In life tables, type II curve shows
   - (A) Low death rate of young
   - (B) High mortality in old
   - (C) Both A and B
   - (D) **Uniform mortality of young and old**

123. In life tables, type III curve shows
   - (A) **Low death rate of young**
   - (B) High mortality in old
   - (C) Both A and B
   - (D) Uniform mortality of young and old

124. The concept of Food web was given by
   - (A) Haeckel
   - (B) Tansley
   - (C) E. Odum
   - (D) **Charles Elton**

125. The Life table for natural insect population was studied in
   - (A) *Helicoverpa armigera*
   - (B) ***Choristoneura fumiferana***
   - (C) *Plodia Interpunctella*
   - (D) None of the above

126. The Life table for natural insect population was studied in
   - (A) *Helicoverpa armigera*
   - (B) ***Choristoneura fumiferana***
   - (C) *Plodia Interpunctella*
   - (D) None of the above

127. The word "Survey" is derived from
   (A) Greek (B) **Latin**
   (C) French (D) English

128. The term "Survey" indicates
   (A) 'A general view (B) An Inspection
   (C) Collection of data/mapping (D) **All of the above**

129. The assessment of insect-pest population and their damage is known as
   (A) **Survey** (B) Surveillance
   (C) Both A and B (D) None of the above

130. The objectives of Survey and surveillance is
   (A) To identify insect-pests and biotypes
   (B) To assess pest damaging levels, EIL, ETLs
   (C) To study the changing status of pests
   (D) **All of the above**

131. The repeated or sequential survey of the same place or locality for some observation to ascertain the changes or fluctuations in the object of study is known as
   (A) Survey (B) **Surveillance**
   (C) Both A and B (D) None of the above

132. The Survey involving the identification of the different species present over an area is termed as
   (A) **Qualitative survey** (B) Quantitative survey
   (C) Collection of data/mapping (D) All of the above

133. The survey involving the estimation of population of one or more species of insects is termed as
   (A) Qualitative survey (B) **Quantitative survey**
   (C) Collection of data/mapping (D) All of the above

134. The sampling methods are used to estimate the density of insects per unit area is termed as
   (A) **Absolute method** (B) Relative method
   (C) Recapture method (D) All of the above

135. The sampling methods are used to indicate of insect abundance or damage relative to different times or locations is termed as
   (A) Absolute method (B) **Relative method**
   (C) Recapture method (D) All of the above

136. The type of Absolute sampling methods are

(A) Unit habitat method (B) Removal trapping

(C) Recapture method (D) **All of the above**

137.The method used in relative sampling are

(A) Plant damage (B) Visual searches

(C) Use of light traps (D) **All of the above**

138. Which sampling method gives accurate information of pest density and damage

(A) **Absolute method** (B) Relative method

(C) Recapture method (D) All of the above

139. The sampling technique where the sample is taken at random with good field coverage todetermine insect numbers or damage per sample unitis

(A) sequential sampling (B) **Random Sampling**

(C) Stratified random sampling (D) All of the above

140. The sampling technique where continuous sampling is done until a pre-established upper or lower infestation level is found.

(A) **Sequential sampling** (B) Random Sampling

(C) Stratified random sampling (D) All of the above

141. The sampling technique involving the division of population into different strata based on distribution of population is

(A) Sequential sampling (B) Random Sampling

(C) **Stratified random sampling** (D) All of the above

142. The Nodal agency for pest surveillance programme in India is

(A) **DPPQ&S** (B) NCIPM

(C) ICAR (D) None of the above

143. The techniques for pest surveillance is

(A) Fixed plot survey (B) Roving Survey

(C) **Both A and B** (D) None of the above

144. The repeated observations at suitable interval (1-2 weeks)from the same fields selected in the beginning of surveillance workis

(A) **Fixed plot survey** (B) Roving Survey

(C) Both A and B (D) None of the above

145. The survey is conducted every week in two fields in each ofthe four villages randomly selectedis

(A) Fixed plot survey (B) **Roving Survey**

(C) Both A and B (D) None of the above

146.The advantage of Remote sensing is

(A) Synoptic view (B) Repetitivity

(C) Coverage of inaccessible areas (D) **All of the above**

147. The White card report prepared after surveillance indicate

(A) **Pest is under control** (B) Pest is noticed at 50% of RTL

(C) Pest reached the ETL (D) Allof the above

148. The Yellow card report prepared after surveillance indicate

(A) Pest is under control (B) **Pest is noticed at 50% of RTL**

(C) Pest reached the ETL (D) Allof the above

149. The Red card report prepared after surveillance indicate

(A) Pest is under control (B) Pest is noticed at 50% of RTL

(C) **Pest reached the ETL** (D) All of the above

150. The main role of Forecasting is

(A) to predict the forthcoming infestation level of the pest

(B) to find out the critical stage for pest management

(C) **Both A and B**

(D) None of the above

151. The species where members of the same generation live together, but without cooperative brood care are termed as

(A) **Communal** (B) Eusocial

(C) Semisocial (D) Quasisocial

152. The species engaged in cooperative brood care, have reproductive castes and overlapping generations. are termed as

(A) Communal (B) **Eusocial**

(C) Semisocial (D) Quasisocial

153. The species where members of the same generation live together and cooperate in brood care are termed as

(A) Communal (B) Eusocial

(C) Semisocial (D) **Quasisocial**

154. The species where there is cooperative brood care of females of the same generation as well as reproductive caste differentiation are termed as

(A) Communal (B) Eusocial

(C) **Semisocial** (D) Quasisocial

155. The phenomenon of Communalism is observed in

(A) *Nicophorus* burying beetle (B) **Psocoptera**

(C) Honeybees (D) *Polistes* wasps

156. The phenomenon of Semisociality is present in

(A) *Nicophorus* burying beetle (B) Psocoptera

(C) Honeybees (D) ***Polistes* wasps**

157. The phenomenon of Quasisociality is present in

(A) ***Nicophorus* burying beetle** (B) Psocoptera

(C) Honeybees (D) *Polistes* wasps

158. The phenomenon of Eusociality is observed in

(A) *Nicophorus* burying beetle (B) Psocoptera

(C) **Honeybees** (D) *Polistes* wasps

159. The insects having Eusociality is

(A) Ants (B) Termites

(C) Honeybees (D) **All of the above**

160. The existence of two or more forms of individuals with the same insect species is termed as

(A) Polyphenism (B) **Polymorphism**

(C) Morphsism (D) Polyethism

161. The occurrence of different phenotypes within the same insect species is termed as

(A) **Polyphenism** (B) Polymorphism

(C) Morphsism (D) Polyethism

162. The Computer-based Climatic model "CLIMEX" for determining arthropod distribution and abundance was developed by

(A) Tansley (B) **R.W. Sutherst and G.F. Maywald**

(C) Bower (D) R.J. Cowley

163. The Dufour's Glands are present in

(A) **Ants** (B) **Honeybee**

(C) Earwigs (D) Termites

164. The phenonmenon by which the Insects blend in with their surroundings and often manage to escape detection by enemies is

(A) Mimesis (B) **Crypsis**

(C) Aposematism (D) Mimicry

165. The mimicry of natural objects is often known as

(A) **Mimesis** (B) Crypsis

(C) Aposematism (D) Mimicry

166. The phenomenon of "advertising" signal in the form of bright warning colourations

(A) Mimesis (B) Crypsis

(C) **Aposematism** (D) Mimicry

167. The type of mimicry displayed in Viceroy butterflies

(A) **Mullerian Mimicry** (B) Batesian mimicry

(C) Both A and B (D) Non of the above

168. The Dufour's Glands are present in

(A) **Ants** (B) Honeybee

(C) Earwigs (D) Termites

169. In an ecosystem, the energy transferred from one trophic level to another is

(A) 90% (B) 50%

(C) **10%** (D) 1%

Unit-5

Insect Pest Management

Q 1. Match the following

1. The book "Introduction to Insect Pest Management" is authored by

(A) RL Pedigo (B) **RL Metcalf and WH Luckman**

(C) D Dent (D) AR Horowitz and I Ishaya

2. The author of "Entomology and Pest Management" is

(A) **RL Pedigo** (B) RL Metcalf and WH Luckman

(C) D Dent (D) AR Horowitz and I Ishaya

3. The book " Insect Pest Management" is authored by

(A) RL Pedigo (B) RL Metcalf and WH Luckman

(C) **D Dent** (D) AR Horowitz and I Ishaya

4. The book "Insect Pest Management: Field and Protected crops" is authored by

(A) RL Pedigo (B) RL Metcalf and WH Luckman

(C) D Dent (D) **AR Horowitz and I Ishaya**

5. The insecticidal properties of DDT was discovered by by

(A) Geigy (B) M. Faraday

(C) Gerhard Schrader (D) **Paul Muller**

6. The Concept of Economic Threshold Level was introduced by

(A) RL Pedigo (B) **Stern**

(C) Dent (D) Smith

7. The term 'Integrated Control' was coined by

(A) **Gierer and Clark** (B) Stern

(C) Dent (D) Smith

8. The term 'Integrated Pest Management' was coined by

(A) Giererand Clark (B) Stern

(C) Dent (D) **Smith and Bosch**

9. The term 'Integrated Pest Management' was coined by Smith and Bosch in the year

(A) 1951 (B) **1967**
(C) 1975 (D) 1985

10. The concept of Environment Injury level' was given by

(A) Giererand Clark (B) **L. P. Pedigo and L.G. Highley**
(C) Dent (D) Smith and Bosch

11. The term "Pest'' is derived from

(A) French (B) Latin
(C) **Both A and B** (D) Italic

12. In India, the first IPM programme 'Operational Research Project was launched in

(A) **1974-75** (B) 1984-85
(C) 1991-92 (D) 1999-2000

13. The Nodal agency entrusted with implementation of IPM programme in the Country is

(A) **DPPQ & S** (B) NCIPM
(C) ICAR (D) NIPHM

14. The National Agriculture Policy envisages IPM as the main strategy of plant protection was in

(A) 1975 (B) 1985
(C) 1991 (D) **2001**

15. An increase in population of target insect after insecticidal spray is referred to as

(A) Resistance (B) **Resurgence**
(C) Secondary pest outbreak (D) Replacement

16. The Pest occur regularly and serious manner called

(A) **Major** (B) Sporadic
(C) Epidemic (D) Endemic

17. The National Institute of Plant Health Management (NIPHM) is located at

(A) Coimbatore (B) New Delhi
(C) Bangalore (D) **Hyderabad**

18. The National Centre of Insect Pest Management is located at

(A) Coimbatore (B) **New Delhi**
(C) Bangalore (D) Hyderabad

19.The National Bureau of Agricultural Insect Resources (NBAIR) is located at

(A) Coimbatore (B) New Delhi

(C) **Bangalore** (D) Hyderabad

20. The Destructive Insect Pest Act was enacted during

(A) 1951 (B) **1914**

(C) 1946 (D) 1968

21. The Insecticide Act was enacted during

(A) 1951 (B) 1914

(C) 1946 (D) **1968**

22. The Insecticide Rules were implemented during

(A) 1951 (B) 1914

(C) **1971** (D) 1968

23. The Silent Spring was written by

(A) **RacahelCarson** (B) Racahel Raymond

(C) Metcalf (D) S. Pradhan

24. Rachel Carson's book which created awareness about ill effects of pesticides

(A) **Silent Spring (1962)** (B) Pesticides-Boon or Vane

(C) Falling Spring (1962) (D) None of the above

25. Rachael carson's *Silent Spring* (1962) pointed which insecticide group responsible for thinning of egg shell or reproductive failure in birds

(A) **Organochlorines** (B) Organophosphates

(C) Carbamates (D) All the above

26.The organochlorines are responsible for

(A) Bio-concentration (B) Bioaccumulation

(C) Bio-magnification (D) **All the above**

27. Accumulation of insecticides at different trophic levels of food chain

(A) Bio-concentration (B) Bioaccumulation

(C) Bio-magnification (D) **All the above**

28. The per cent of the pesticide applied to crop reaches the target.

(A) **1 per cent** (B) 10 per cent

(C) 15 per cent (D) 50 per cent

29. An increase in population of target insect after insecticidal spray is referred to as

(A) **Resurgence** (B) Hormoligosis

(C) Replacement (D) Secondary pest outbreak

30. Situation where a population, after having been suppressed, rebounds to number greater than before suppression occurred

(A) **Resurgence** (B) Hormoligosis

(C) Replacement (D) Secondary pest outbreak

31. Situation when a major pest is suppressed and continues to be suppressed by a tactic but is replaced but is replaced by another pest, previously with minor status

(A) Replacement (B) Secondary pest outbreak

(C) **Both A and B** (D) Hormoligosis

32. Direct favourable influence of pesticides on physiology and behaviour of insect pest is known as

(A) Resurgence (B) **Hormoligosis**

(C) Replacement (D) Secondary pest outbreak

33. The Increased resistance leads to vicious cycle of increased dosage and frequency of pesticide application

(A) **Pesticide treadmill** (B) Excessive pesticide use

(C) Pesticide Overdose (D) All of the above

34. The insect -pests causing damage ranging 5-10 per cent are termed as

(A) **Minor pest** (B) Major pest

(C) Key pest (D) Regular pest

35. The insect -pests causing damage >10 per cent are termed as

(A) Minor pest (B) **Major pest**

(C) Key pest (D) Regular pest

36. The insect-pests ooccurring most frequently in a crop and have close association with that particular crop

(A) Occasional pest (B) Sporadicpest

(C) Key pest (D) **Regular pest**

37. The insect-pests ooccurring infrequently in a crop and do not have close association with that particular crop

(A) **Occasional pest** (B) Sporadicpest

C) Key pest (D) Regular pest

38. The insect-pests ooccurring almost throughout the year on a crop
    (A) Occasional pest (B) Sporadicpest
    (C) **Persistent pest** (D) Regular pest

39. The insect-pests occurring in a severe form in a region or at a particular season or time
    (A) **Epidemic pest** (B) Endemicpest
    (C) Pandemic pest (D) Key pest

40. The insect-pests occurring regularly and confined to a particular region
    (A) Epidemic pest (B) **Endemic pest**
    (C) Pandemic pest (D) Key pest

41. The insect-pests having GEP above the EIL and require regular and constant interventions to produce marketable crops
    (A) Epidemic pest (B) Major pest
    (C) Minor pest (D) **Key pest**

42. The insect-pests having GEP very close to EIL or coincides with the EIL
    (A) Epidemic pest (B) **Major pest**
    (C) Minor pest (D) Key pest

43. The insect-pests having GEP below the EIL and rarely crosses the EIL
    (A) Epidemic pest (B) Major pest
    (C) **Minor pest** (D) Key pest

44. The insect-pests designated as national pest
    (A) Root knot nematode (B) Diamond back moth
    (C) *Helicoverpa armigera* (D) **All of the above**

45. Which of the following is a monophagous pest
    (A) Root knot nematode (B) Diamond back moth
    (C) *Helicoverpa armigera* (D) **Rice stem borer**

46. Which of the following is a monophagous pest
    (A) Brinjal shoot and fruit borer (B) Rice stem borer
    (C) Mango mealy bug (D) **All of the above**

47. Which of the pest outbreak was observed in Punjab (2015)
    (A) **Whitefly on cotton** (B) *Helicoverpa* on cotton
    (C) Mealy bug on cotton (D) Aphid on mung bean

48. In Paddy, high dose of Nitrogen flare up the incidence of

(A) Leaf folder (B) Yellow stem borer
(C) BPH (D) **All the above**

49. In Paddy, closer spacing flare up the incidence of

(A) Leaf folder (B) BPH
(C) GLH (D) **All the above**

50. The fertilizer promoting resistance against pest attack is

(A) Ammonium sulfate (B) Urea
(C) **MOP** (D) DAP

51. The fertilizer promoting resistance against pest attack is

(A) Ammonium sulfate (B) Urea
(C) **MOP** (D) DAP

52. The pest management programme should be

(A) Comprehensive for the production system
(B) Selective for the pest
(C) Compatible with ecological principles
(D) **All the above**

53. The concept of IPM was developed due to

(A) Development of insecticide resistance
(B) Resurgence
(C) Environmental contamination
(D) **All the above**

54. The objectives of pest management is

(A) To reduce pest status below economic injury level
(B) To use eco-friendly methods
(C) To apply control measures only when needed
(D) **All the above**

55. The aim of IPM is

(A) Complete elimination of the pest may not be feasible nor even desirable
(B) Treat the crop when necessary
(C) **Both A and B**
(D) None of the above

56. The requirements for successful pest management programme
   (A) Correct identification and life history of insect pests
   (B) pest surveillance and forecasting
   (C) Selection of suitable method
   (D) **All the above**

57. The concept of Gain threshold was given by
   (A) Stern et al. (B) Smith
   (C) **Stone and Pedigo** (D) All the above

58. The concept of Economic threshold was given by
   (A) Stern et al. (B) Smith
   (C) Stone and Pedigo (D) **F. DuToit**

59. The amount of injury which will justify the cost of artificial control
   (A) EIL (B) ETL
   (C) **Economic damage** (D) Gain threshold

60. The effect of pest activity on host physiology that is usually harmful
   (A) Damage (B) **Injury**
   (C) Economic damage (D) Gain threshold

61. The measurable loss of yield or quality is termed as
   (A) **Damage** (B) Injury
   (C) Economic damage (D) Gain threshold

62. The average population density of insect over a long period of time
   (A) Damage (B) Injury
   (C) **GEP** (D) Gain threshold

63. The lowest level of injury where damage can be measured is termed as
   (A) **Damage boundary** (B) Injury
   (C) Economic damage (D) Gain threshold

64. The lowest level of injury where damage can be measured is termed as
   (A) **Damage boundary** (B) Injury
   (C) Economic damage (D) Gain threshold

65. The Lowest population at which the pest will cause economic damage is termed as
   (A) **Economic Injury Level** (B) Economic Threshold Level
   (C) Economic damage (D) Gain threshold

66. The Economic Injury Level is expressed as

(A) **No. of insects/area** (B) Per cent damage
(C) Both A and B (D) None of the above

67. The population density at which control measures should be applied to prevent increasing pest population

(A) EconomicInjuryLevel (B) **Economic Threshold Level**
(C) Economic damage (D) Gain threshold

68. The ETL always represent a pest density

(A) **Lower than EIL** (B) Above ETL
(C) Equal to EIL (D) Equal or lower than ETL

69. The Pedigo method for determination of EIL is

(A) Q = C/PDK (B) **EIL = C/VIDK**
(C) EIL = C/PDK (D) Q = C/VIDK

70. The Norton method for determination of EIL is

(A) **Q = C/PDK** (B) EIL = C/VIDK
(C) EIL = C/PDK (D) Q = C/VIDK

71. In Pedigo method for determination of EIL, K is

(A) Reduction in pest attack (B) Loss in yield
(C) **Proportion reduction in pest damage** (D) Reduction in yield

72. The EIL is

(A) Directly proportional to cost of management
(B) Market value of commodity
(C) Injury units per insect
(D) **All of the above**

73. The ETL value for Yellow stem borer of rice is

(A) 5% dead hearts (B) 5% white ears
(C) 1 egg mass/m^2 (D) **All the above**

74. The ETL value for Brown Plant Hopper of rice is

(A) **5-10 hopper/hill** (B) 10-20 hopper/hill
(C) 2-5 hopper/hill (D) 25-35 hopper/hill

75. The ETL value for Maize stem borer is

(A) **5-10% infestation** (B) 5% dead hearts
(C) 10-15% dead hearts (D) 15% dead hearts

76. The ETL value for Mustard aphid is

(A) 20-30 aphids/10cm terminal shoot

(B) 30-50 aphids/10cm terminal shoot

(C) **50-60 aphids/10cm terminal shoot**

(D) 60-75aphids/10cm terminal shoot

77. The ETL value for brinjal shoot and fruit borer is

(A) 1-5% shoot infestation (B) 1-5% fruit infestation

(C) **Both A and B** (D) 5-10% shoot infestation

78. The ETL value for Tomato fruit borer is

(A) 1-5% shoot infestation (B) **1-5% fruit infestation**

(C) Both A and B (D) 5-10% shoot infestation

79. The ETL value for pink bollworm in cotton is

(A) 1-5% boll infestation (B) **5-10% fruit infestation**

(C) 10-15% boll infestation (D) >15% shoot infestation

80. The ETL value for American bollworm in cotton is

(A) 1-5% boll infestation (B) **5-10% fruit infestation**

(C) 10-15% boll infestation (D) >15% shoot infestation

81. The ETL value for whitefly in cotton is

(A) 2-5 adults/leaf (B) **8-10 adults/leaf**

(C) 10-15 adults/leaf (D) >15adults/leaf

82. The ETL value for Sugarcane pyrilla is

(A) **3-5 nymphs or adults/leaf** (B) 8-10 nymphs or adults/leaf

(C) 10-15 nymphs or adults/leaf (D) >15nymphs or adults/leaf

83. The ETL value for codling moth in apple is

(A) 5-10% incidence (B) 4-8% incidence

(C) 2-4 % incidence (D) **1-2% incidence**

84. The potential of IPM includes

(A) Sustainable (B) Economical

(C) Maintain environmental quality (D) **All of the above**

85. The major limitations in implementations are

(A) Lack of interdisciplinary approach

(B) Lack of IPM information among farmers

(C) Direct conflict between IPM objectives and pesticide industry

(D) **All of the above**

86. Which of the following is not a reason for pest out break
   (A) Afforestation
   (B) Accidental introduction of insect species
   (C) Intensive and extensive cultivation
   (D) **All of the above**

87. It is advised to initiate management practice when the pest reaches
   (A) EIL (B) DB
   (C) **ETL** (D) GEP

88. Is EIL of a pest constant at all times in all places
   (A) Yes (B) **No**
   (C) Can't say (D) Don't Know

89. Which of the following method is best suited for surveying a large area in short period
   (A) **Roving survey** (B) Fixed plot survey
   (C) Both A and B (D) None of the above

90. In which of the following type of pest GEP is always above the EIL
   (A) Potential pest (B) **Key pest**
   (C) Sporadic pest (D) Major pest

91. Agricultural ecosystem is more diverse than the forest ecosystem
   (A) Yes (B) **No**
   (C) Equally diverse (D) Can't say

92. Which of the following is not a physical means of pest management
   (A) Use of light traps (B) Coating of stored grains with oils
   (C) Trenching (D) **Alternate wetting and drying of field**

93. Leaving pest residues is one of the concept of
   (A) Integrated crop management
   (B) Integrated resistance management
   (C) **Integrated pest management**
   (D) Integrated pest control

94. Manipulating the date of sowing is an important method of pest management
   (A) **Cultural** (B) Mechanical
   (C) Physical (D) Natural

95. Hand picking and shaking of branches of plants comes under the following method of pest control

(A) Cultural (B) **Mechanical**

(C) Physical (D) Natural

96. Which of the following predators was introduced into India from Austarlia for the management of citrus mealy bugs In 1898

(A) *Oecophylla smaragdina* (B) ***Cryptolaemus montrozirie***

(C) *Rodolia cardinalis* (D) *Spogossia bezziana*

97. All Regular pests are Epidemic pests 2. All Epidemic pests are not Regular pests

(A) First statement is true but not the second statement

(B) **Second statement is true but not the first statement**

(C) Both the statements are true

(D) Both the statements are false

98. The minimum insect population that produces incremental damage equal to the cost of protection is referred as

(A) **EIL** (B) GEP

(C) ETL (D) Modified equilibrium position

99. Which of the following do not influence EIL

(A) Market value of the crop (B) Management cost

(C) Crop susceptibility (D) **Economic status of the farmer**

100. Manipulation of abiotic factors to reduce the pest status is known as

(A) **Physical method** (B) Mechanical method

(C) Cultural method (D) All of the above

101. Alternate wetting and drying of paddy field help to reduce the status of

(A) **Brown plant hopper** (B) Green Leaf Hopper

(C) Yellow stem borer (D) All of the above

102. The moisture per cent recommended for storage of food grains is

(A) **<10%** (B) 10-15%

(C) 15-20% (D) None of the above

103. Grains stored at around 10% moisture content escape from the attack of insect-pests except

(A) *Lasioderma serricorne* (B) *Sitophilus oryzae*

(C) **Khapra beetle** (D) All of these

104. Light traps are use to reduce the pest status of
(A) White grub (B) Paddy Leaf folder
(C) Hairy caterpillar (D) **All of the above**

105. Yellow sticky traps are use to reduce the pest status of
(A) Aphids (B) Whiteflies
(C) Jassids (D) **All of the above**

106. Sterile male release technique was done by Knipling to control screwworm fly in
(A) Madagascar (B) Kenya
(C) **Curacao Island** (D) Mexico

107. The radiation used for Sterile male release technique by Knipling to control screwworm fly is
(A) **Gamma radiation (Co60)** (B) Gamma radiation (Po50)
(C) Gamma radiation (U60) (D) None of the above

108. Sterile male release technique was done by Knipling to control
(A) **Cattle screw worm fly** (B) Gad fly
(C) Tabanid fly (D) All of the above

109. The porous finely divided silica gel used against stored pest is
(A) **Drie-Die** (B) Red earth
(C) Activated clay (D) All of the above

110. Hook method (use of Iron hook) in coconut is used to control
(A) Red palm weevil (B) Black head caterpillar
(C) **Rhinoceros beetle** (D) All of the above

111. Rope method, use of coir rope soaked in one part of kerosene and one part of water in rice crop is used to control
(A) Yellow stem borer (B) **Leaf folder**
(C) Gundhi bug (D) All of the above

112. Clipping of leaves tip in nursery in rice crop is used to control
(A) **Yellow stem borer** (B) Leaf folder
(C) Gundhi bug (D) All of the above

113. Clipping and destruction of infested twigs (top 10-15cm) in mustard help to reduce the infestation of
(A) Mustard sawfly (B) Cabbage butterfly
(C) Painted bug (D) **Mustard aphid**

114. Which one of the following pests of mango can be controlled by grease banding around tree trunk

(A) **Mealybug** (B) Tree borer

(C) Hoppers (D) Blossom midge

115. Bagging of fruits in pomegranate is done to avoid the attack of

(A) *Otheris materna* (B) *Scelodonta strigicollis*

(C) *Papilio demoleus* (D) ***Virachola isocartes***

116. Bagging of fruits in Citrus is done to avoid the attack of

(A) ***Otheris materna*** (B) *Scelodonta strigicollis*

(C) *Papilio demoleus* (D) *Virachola isocartes*

117. Splitting of pulses avoid the attack of

(A) **Pulse beetle** (B) Pod borer

(C) Both A and B (D) None of the above

118. Trenching in wheat is done to avoid the attack of

(A) **Army worm** (B) *Helicoverpa armigera*

(C) Termites (D) Ghujhia weevil

119. Fish meal traps are used in sorghum to avoid the attack of

(A) *Chilo partellus* (B) ***Atherigona soccata***

(C) *Perigrinus maidis* (D) *Contarinia sorghicola*

120. Tilling of soil near bunds or reducing the bund size in paddy is done to reduce the attack of

(A) Leaf folder (B) Gall midge

(C) Hispa **(D) Grasshopper**

121. Fall ploughing in crops is recommended to reduce the attack of

(A) Root grub (B) *Helicovera armigera*

(C) Red hairy caterpillar (D) **All of the above**

122. Frequent raking and hoeing of soil is recommended to reduce the attack of

(A) Fruit flies (B) Red pumpkin beetle

(C) **Both A and B** (D) None of the above

123. Frequent raking and hoeing of soil is recommended to reduce the attack of

(A) Fruit flies (B) Red pumpkin beetle

(C) **Both A and B** (D) None of the above

124. Removal of alternate host help to reduce the attack of

(A) ***Helicoverpa armigera*** (B) *Spilosoma obliqua*

(C) *Spodoptera litura* (D) None of the above

125. Planting small area of a crop or other species near or in the target crop is known as

(A) Intercropping (B) **Trap cropping**

(C) Mixed cropping (D) None of the above

126. Trap crop recommended for diamond back moth in cauliflower is

(A) **Mustard** (B) Castor

(C) Marigold (D) Cucumber

127. Trap crop recommended for *Helicoverpa armigera* management in Tomato is

(A) Mustard (B) Castor

(C) **Marigold** (D) Cucumber

128. Trap crop recommended for *Chilo partellus* in maize is

(A) Mustard (B) Castor

(C) Marigold (D) **Sorghum**

129. Trap crop recommended for Bollworms in cotton is

(A) **Okra** (B) Castor

(C) Marigold (D) Cucumber

130. Trap crop recommended for *Spodoptera litura* in groundnut is

(A) Okra (B) **Castor**

(C) Marigold (D) Cucumber

131. Strategy recommended for controlling *aphid* transmitting non persistent viruses is

(A) Mixed crop (B) **Barrier crops**

(C) Inter crops (D) Trap crops

132. Trap crop recommended for reducing thrips infestation in cotton is

(A) Garlic (B) Onion

(C) **Both A and B** (D) Castor

133. Intercropping help in reducing the pest population due to

(A) Availability of alternate prey

(B) Decreased colonization and reproduction in pests

(C) Chemical repellency, masking, feeding inhibition

(D) **All of the above**

134. Intercrop recommended for managing Early shoot borer in sugarcane is

(A) **Green gram** (B) Sorghum

(C) Mustard (D) Red gram

135. Intercrop recommended for managing Ear head bug in Sorghum is

(A) Green gram (B) Cowpea

(C) Mustard (D) **Red gram**

136. Intercrop recommended for managing Gram pod borer in Chickpea is

(A) Wheat (B) Safflower

(C) Mustard (D) **All the above**

137. Intercrop recommended for managing stem borer in Sorghum is

(A) Green gram (B) **Cowpea**

(C) Mustard (D) Redgram

138. The high dose of Nitrogen in wheat increase the attack of

(A) Armyworm (B) Aphids

(C) *Helicoverpa* (D) **Cutworm**

139. The Early planting (upto 3rd week of June) of rice avoid the attack of

(A) Gall midge (B) **Yellow stem borer**

(C) Shoot fly (D) BPH

140. The planting (end of July) of rice avoid the attack of

(A) Gall midge (B) Yellow stem borer

(C) Shoot fly (D) **BPH**

141. The delayed sowing in sorghum increase the incidence of

(A) Shoot fly (B) Shoot bug

(C) **Leafhopper** (D) Stem borer

142. For every 10 days delay in sowing increase the pod damage of gram pod borer by

(A) **4.02%** (B) 8.2%

(C) 10.2% (D) 20%

143. Higher seed rate in sugarcane lower the incidence of

(A) Zig zag borer (B) Early shoot borer

(C) Gurdaspur borer (D) **Top shoot borer**

144. Dense planting in rice increase the incidence of

(A) Leaf folder (B) BPH

(C) **Both A and B** (D) None of the above

145. Less plant population in chickpea increase the incidence of

(A) **Aphids** (B) *Helicoverpa armigera*

(C) Both A and B (D) None of the above

146. Destruction of alternate weeds in rice lower the incidence of

(A) Gall midge (B) WBPH

(C) GLH (D) **All of the above**

147. Destruction of alternate weeds in Sorghum lower the incidence of

(A) Shoot fly (B) Shoot bug

(C) **Ear head bug** (D) Stem borer

148. Destruction of alternate weed, *Achyranthus aspera* in groundnut lower the incidence of

(A) Whitefly (B) Aphids

(C) **Thrips** (D) All of the above

149. Copious irrigation in fruit tree nurseries lower the incidence of

(A) **Termites** (B) Aphids

(C) Thrips (D) All of the above

150. Severe pest outbreaks are mostly observed in

(A) Overlapping cropping seasons (B) mono-cropping

(C) Cultivation of same variety (D) **All of the above**

151. High silica content in rice reduce the incidence of

(A) **Yellow stem borer** (B) Leaf folder

(C) BPH (D) All of the above

152. Roughness and toughness of seed coat in gram reduce the incidence of

(A) **Pulse beetle** (B) *Helicoverpa*

(C) Pod fly (D) All of the above

153. The presence of purple foliage and apetalous flowers in mustard reduce the incidence of

(A) Sawfly (B) **Aphid**

(C) Painted bug (D) All of the above

154. Okra leaves type cultivars in Cotton are resistant to

(A) American bollworm (B) Pink bollworm

(C) Whiteflies (D) **All of the above**

155. The skin toughness and pulp hardiness of fruit in brinjal reduce the incidence of
(A) Stem borer (B) **Shoot and fruit borer**
(C) Hadda beetle (D) All of the above

156. The presence of Low amino acid and sugars in rice reduce the incidence of
(A) **Yellow stem borer** (B) Leaf folder
(C) BPH (D) All of the above

157. The Low asparagine content in rice reduce the incidence of
(A) Yellow stem borer (B) Leaf folder
(C) **BPH** (D) All of the above

158. The Low sugars and high glucosinolates in mustard reduce the incidence of
(A) Sawfly (B) **Aphid**
(C) Painted bug (D) All of the above

159. The presence of DIMBOA and MBOA in maize reduce the incidence of
(A) Western corn rootworm (B) European corn borer
(C) **Both A and B** (D) None of the above

160. The DIMBOA in maize
(A) **Reduce digestibility** (B) Reduce efficiency of digested food
(C) Both A and B (D) None of the above

161. The MBOA in maize
(A) Reduce digestibility (B) **Reduce efficiency of digested food**
(C) Both A and B (D) None of the above

162. The lethal silk factor in maize is
(A) DIMBOA (B) MBOA
(C) **Maysin** (D) None of the above

163. The lethal silk factor (Maysin) in maize reduce the incidence of
(A) Corn earworm (B) Fall armyworm
(C) **Both A and B** (D) None of the above

164. The cultivar shows resistance against pigeon pea pod borer
(A) Abhaya (B) Bahar
(C) Bori (D) **All of the above**

165. The cultivar shows resistance against Brinjal shoot and fruit borer
(A) Pusa Purple Long (B) Punjab neelam
(C) Punjab Barsati (D) **All of the above**

166. The cultivar shows resistance against Wooly apple aphid
 (A) Northern Spy (B) Golden Delicious
 (C) Norton Rootsock 778 (D) **All of the above**

167. The cultivar shows resistance against yellow stem borer in rice
 (A) Vikas (B) Ratna
 (C) Saket (D) **All of the above**

168. The cultivar shows resistance against Onion thrips
 (A) Agrifound Light Red (B) Pusa Red
 (C) Agrifound Dark Red (D) **All of the above**

169. The cultivar shows resistance against Chillies thrips
 (A) Pusa Jwala (B) NP-46
 (C) CH-1 (D) **All of the above**

170. The cultivar shows resistance against Tomato fruit borer
 (A) Azad Pusa Hybrid 4 (B) Punjab Keasri
 (C) Pant Bahar (D) **All of the above**

Unit-6

Host Plant Resistance

MCQ's

1. Father of Host Plant Resistance is

(A) Knipling (B) Snodgrass

(C) Hagstrum (D) **Painter RH**

2. Host Plant Resistance can be assessed by

(A) Resistance is heritable (B) It is relative

(C) Resistance is variable (D) **All the above**

3. The variety which a specific insect will never injure under any known conditions

(A) **Immune** (B) Resistant

(C) Susceptible (D) Tolerant

4. The variety with high resistance is one which possess qualities resulting in a small damage by a specific pest under a given set of conditions

(A) Immune (B) **Resistant**

(C) Susceptible (D) Tolerant

5. The variety which shows average or more than average damage caused by the insects

(A) Immune (B) Resistant

(C) **Susceptible** (D) Tolerant

6. The resistance resulting from the transitory characters in potentially susceptible host plants

(A) Ecological resistance (B) Apparent resistance

(C) Pseudo-resistance (D) **All the above**

7. The example of ecological resistance

(A) Host evasion (B) Induced resistance

(C) Host escape (D) **All of the above**

8. The ecological resistance in which host may pass through the most susceptible stage quickly or at times when number of insects is less

(A) **Host evasion** (B) Induced resistance
(C) Host escape (D) All of the above

9. The Increase in resistance temporarily as a result of some changed conditions of plant or environment such as change in the amount of water or nutrient status of soil

(A) Host evasion (B) **Induced resistance**
(C) Host escape (D) All of the above

10. The Absence of infestation or injury to the host plant because of transitory circumstances such as incomplete infestation

(A) Host evasion (B) Induced resistance
(C) **Host escape** (D) All of the above

11. The heritable qualities possessed by an organism which influence the ultimate degree of damage done by a insect species having a prior continuous, co-evolutionary history with that species of organism

(A) **Sympatric resistance** (B) Allopatric resistance
(C) Evolutionary resistance (D) All the above

12. The heritable qualities possessed by an organism which influence the ultimate degree of damage done by a insect species having no prior continuous, co-evolutionary history with that species of organism

(A) Sympatric resistance (B) **Allopatric resistance**
(C) Evolutionary resistance (D) All the above

13. The sympatric resistance is governed by

(A) **Major genes** (B) Polygenes
(C) Oligogenes (D) None of the above

14. The Allopatric resistance is governed by

(A) Major genes (B) **Polygenes**
(C) Oligogenes (D) None of the above

15. The resistance in Plant through physical means or through production of chemicals or both is known as

(A) **Intrinsic resistance** (B) Extrinsic resistance
(C) Allopatric resistance (D) None of the above

16. The resistance natural enemies (third trophic level) of insect-pests (second trophic level) benefit the host plants (first trophic level) by reducing the pest incidence

(A) Intrinsic resistance (B) **Extrinsic resistance**

(C) Allopatric resistance (D) None of the above

17. The resistance against western corn rootworm is governed by

(A) **Single gene** (B) Few genes

(C) Many genes (D) None of the above

18. The resistance against Hessian fly resistance in wheat is governed by

(A) Single gene (B) **Few genes**

(C) Many genes (D) None of the above

19. The Resistance to European corn borer in corn is governed by

(A) Single gene (B) Few genes

(C) **Many genes** (D) None of the above

20. The resistance in a variety incorporated against a primary pest, confers resistance to another insect is called as

(A) Pureline resistance (B) **Cross resistance**

(C) Multiple resistance (D) None of the above

21. The resistance in a variety incorporated provide protection to a variety against different environmental stresses is called as

(A) Pureline resistance (B) Cross resistance

(C) **Multiple resistance** (D) None of the above

22. The Token (positive) Stimuli Theory was proposed by

(A) Kennedy and Booth (B) **Fraenkel**

(C) Dethier (D) Thornstein

23. The Negative Stimuli Theory was proposed by

(A) Kennedy and Booth (B) Fraenkel

(C) Dethier (D) **Thornstein and Jeremy**

24. The Nutritional imbalance Theory was proposed by

(A) Kennedy and Booth (B) Fraenkel

(C) Dethier (D) **House**

25. The Botanical instinct Theory was proposed by

(A) Kennedy and Booth (B) **Brues**

(C) Dethier (D) House

26. The Chemosensory basis of host selection Theory was proposed by
    (A) Kennedy and Booth (B) Brues
    (C) **Dethier** (D) House
27. The Appropriate and inappropriate landing theorywas proposed by
    (A) **Finch and Coller** (B) Brues
    (C) Dethier (D) House
28. The Dual discrimination theory was proposed by
    (A) Finch and Coller (B) **Kennedy and Booth**
    (C) Dethier (D) House
29. The **Non preference, antibiosis and tolerance theory** was proposed by
    (A) **Painter** (B) Brues
    (C) Dethier (D) House
30. The non- preference plant resistance to insects is also known as
    (A) Antibiosis (B) **Antixenosis**
    (C) Tolerance (D) Host evasion
31. The host plant is considered by insect-pest as undesirable or bad host in case of
    (A) Non preference (B) Antixenosis
    (C) **Both A and B** (D) Antibiosis
32. The presence of Frego bracts in cotton imparts resistance against
    (A) **Boll weevil** (B) whitefly
    (C) Jassids (D) Thrips
33. The Okra leaves cultivars in cotton imparts resistance against
    (A) Bollworm complex (B) whitefly
    (C) **Both A and B** (D) Thrips
34. The presence of high silica content in sorghum in 4th and 6th leaf stage imparts resistance against
    (A) **Shoot fly** (B) stem weevil
    (C) Stem borer (D) Thrips
35. The high silica content in rice imparts resistance against
    (A) Asiatic stem borer (B) yellow stem borer
    (C) **Both A and B** (D) None of the above

36. The tightness of glume closure in sorghum imparts resistance against
    (A) **Sorghum midge** (B) whitefly
    (C) Jassids (D) Thrips

37. The Purple foliage and apetalous flowers in *Brassica* spp.imparts resistance against
    (A) Mustard sawfly (B) **Mustard aphid**
    (C) Painted bug (D) All of the above

38. The Skin Toughness and pulp hardiness of fruit in brinjal imparts resistance against
    (A) **Shoot and fruit borer** (B) Leaf hopper
    (C) Hadda beetle (D) All of the above

39. The presence of awn in wheat imparts resistance against
    (A) **Aphids** (B) Thrips
    (C) *Helicoverpa armigera* (D) All of the above

40. The Leaf wax (glossy leaves) in *Brassica* spp.imparts resistance against
    (A) Mustard sawfly (B) Mustard aphid
    (C) Painted bug (D) **DBM**

41. The High P, K and Mg and low N and Fe in cotton imparts resistance against
    (A) Jassids (B) Stem weevil
    (C) Whitefly (D) **All of the above**

42. The High Pentadecanal content in rice imparts resistance against
    (A) ***Chilo suppresalis*** (B) *Chilo partellus*
    (C) *Scirpophaga incertulas* (D) All of the above

43. The Low amino acids and sugars in rice imparts resistance against
    (A) *Chilo suppresalis* (B) *Chilo partellus*
    (C) ***Scirpophaga incertulas*** (D) All of the above

44. The Low chlorophyll, carotenoids, N and crude protein content in maize imparts resistance against
    (A) *Chilo suppresalis* (B) ***Chilo partellus***
    (C) *Scirpophaga incertulas* (D) All of the above

45. The High glucosinolates content and low sugar content in *Brassica* imparts resistance against
    (A) **Mustard aphid** (B) Mustard sawfly
    (C) Painted bug (D) All of the above

46. The DIMBOA in maize

(A) **Reduce digestibility** (B) Reduce efficiency of digested food

(C) Both A and B (D) None of the above

47. The MBOA in maize

(A) Reduce digestibility (B) **Reduce efficiency of digested food**

(C) Both A and B (D) None of the above

48. The lethal silk Factor in maize is

(A) DIMBOA (B) MBOA

(C) **Maysin** (D) All of the above

49. The High maysin (a flavonol C-glycoside) content in corn's silk imparts resistance against

(A) Corn earworm (B) armyworm

(C) **Both A and B** (D) All of the above

50. The Low level of total sugar in seeds and pod coat in pigeon pea imparts resistance against

(A) **Pod borer** (B) aphids

(C) Jassids (D) All of the above

51. The Antibiosis phenomenon is due to

(A) Presence of toxic metabolites

(B) Absence or insufficient amount of essential nutrients

(C) Nutrient imbalance

(D) **All of the above**

52. The High gossypol content in gossypol gland density on ovary surface in cotton imparts resistance against

(A) American bollworm (B) Spotted bollworm

(C) Pink bollworm (D) **All of the above**

53. The High Asparigine content in rice imparts resistance against

(A) **Leafhopper** (B) Aphids

(C) mites (D) All of the above

54. The High Phloridzin (Flavonoids) content in apple imparts resistance against

(A) Leafhopper (B) **Aphids**

(C) Mites (D) All of the above

55. The High Phloretin content in apple imparts resistance against

(A) *Empoasca devestans* (B) *Cydia pomonella*

(C) ***Scolytus multistriatus*** (D) All of the above

56. The High Tomatine content in potato imparts resistance against

(A) **BPH** (B) WBPH

(C) GLH (D) All of the above

57. The high Dhurrin content in young leaves in sorghum imparts resistance against

(A) Green bug (B) Migratory locust

(C) Shoot fly (D) **All of the above**

58. The high sinigrin content in young leaves in Crucifers imparts resistance against

(A) Aphids (B) **Cabbage butterfly**

(C) DBM (D) All of the above

59. The Tolerant varieties have

(A) **Higher ETL than susceptible varieties**

(B) Lower ETL than susceptible varieties

(C) Nothing to do with ETL

(D) None of the above

60. The HPR mechanism exerting highest selection pressure

(A) Tolerance (B) Antixenosis

(C) **Antibiosis** (D) All the above

61. The HPR mechanism exerting least selection pressure

(A) **Tolerance** (B) Antixenosis

(C) Antibiosis (D) All the above

62. The biotype development is more in case of

(A) Tolerance (B) Antixenosis

(C) **Antibiosis** (D) All the above

63. Genetic basis of disease resistance was first given by

(A) **Biffen** (B) Stakman

(C) Dethier (D) Thornstein

64. Sinigrin in *Brassica* spp. act as synomone for natural enemy

(A) **Sporadic pest** (B) Regular pest

(C) Occasional pest (D) Key pest

65. The wheat variety resistant to stem sawfly is
(A) **Rescue** (B) Kawvale
(C) Pawnee (D) Mopa

66. The wheat variety resistant to Hessian fly is
(A) Ponca (B) Kawvale
(C) Pawnee (D) **All of the above**

67. The Barley resistant varieties resistant to Green bug is
(A) Dicktoo (B) Will
(C) **Both A and B** (D) None of the above

68. The pest which occurs in an isolated condition is
(A) **Sporadic pest** (B) Regular pest
(C) Occasional pest (D) Key pest

69. Host plant resistance in which the plant passes through a susceptible stage quickly is
(A) Host escape (B) Host tolerant
(C) Host resistance (D) **Host evasion**

70. Which of the following is a type of Ecological Resistance
(A) Host evasion (B) Induced resistance
(C) Escape (D) **All of the above**

71. The nutrition value of the plant and absence of toxic compound help in the process of
(A) Host finding (B) Host recognition
(C) Host acceptance (D) **Host suitability**

72. Theory of chemosensory bases of host selection was proposed by
(A) **Dethier** (B) Painter R H
(C) Morgan (D) Hasgtrum

73. Which of the following is an allomone
(A) **Repellents** (B) Attractant
(C) Excitants (D) All of the above

74. A plant may pass through the most susceptible stage quickly or at time when insect number is less is called
(A) **Host evasion** (B) Induced resistance
(C) Escape (D) All of the above

75. The type of mechanism in which adverse effect of the host plant on the biology of the insects

(A) Antixenosis (B) **Antibiosis**

(C) Non preference (D) Tolerance

76. The number of biotypes in rice brown plant hopper is

(A) **5** (B) 6

(C) 7 (D) 8

77. The number of biotypes in rice gall midge is

(A) 11 (B) 12

(C) **13** (D) 10

78. Maximum number of insect biotypes is observed in

(A) Brown plant hopper (B) Gall midge

(C) **Hessian fly** (D) Pea aphid

79. Which one of the following is an advantage of HPR

(A) Specificity (B) Compatibility

(C) Decrease pesticide application (D) **All of the above**

80. Which one the following is limitation of HPR

(A) Time consuming (B) Biotype development

(C) Genetic limitations (D) **All the above**

81. When was the first transgenic cotton variety approved for commercial cultivation in India

(A) 2000 (B) 2001

(C) **2002** (D) All of the above

82. Bt was first isolated from diseased larvae of

(A) **Silkworm** (B) *Helicoverpa armigera*

(C) Spruce bud worm (D) *Spodoptera litura*

83. The earliest documented case of insect resistance is against

(A) Wheat stem sawfly (B) **Hessian fly**

(C) Sorghum shoot fly (D) Screw Worm Fly

84. Major Gene resistance is also called

(A) Horizontal resistance (B) Oligogenic resistance

(C) **Vertical resistance** (D) Induced resistance

85. DIMBOA is Considered to be resistance factor against European corn borer in
(A) **First generation** (B) Second generation
(C) Third generation (D) Fourth generation

86. A multiple insect and disease resistant variety which is grown on more than 11 milllion ha in the world is
(A) IR8 (B) IR26
(C) **IR36** (D) IR72

87. The Clearance of transgenic cropsfor commercial cultivation in India is given by
(A) **GEAC** (B) RCGM
(C) Both A and B (D) None of the above

88. The ministry deal with the transgenic crops in India is
(A) Ministry of Agriculture
(B) **Ministry of Environment, Forest and Climate Change**
(C) Ministry of Commerce
(D) Ministry of Finance

89. The Organisation entrusted with transgenic crops in India is
(A) **Department of Biotechnology**
(B) Indian Council of Agriculture Research
(C) Council of Scientific and Industrial Research
(D) All the above

90. The first transgenic Mustard cultivar developed in India is
(A) **DMH-11** (B) DCM-11
(C) DUMH-11 (D) Pusa Mustard

91. The transgenic Mustard cultivar developed in India contain
(A) Barnase gene (B) Barstar gene
(C) **Both A and B** (D) None of the above

92. The Barnase and Barstar gene in transgenic Mustard are isolated from
(A) ***Bacillus amyloliquefaciens*** (B) *Bacillus thuringiensis*
(C) *Bacillus subtilis* (D) All the above

93. The function of Barnase gene in transgenic Mustard is
(A) **Impairs pollen production and renders the plant into which it has been introduced male-sterile**
(B) Blocks the action of the "Barstar" gene
(C) Both A and B
(D) None of the above

94. The function of Barstar gene in transgenic Mustard is

(A) Impairs pollen production and renders the plant into which it has been introduced male-sterile

(B) Blocks the action of the "Barnase" gene

(C) Both A and B

(D) None of the above

95. The first transgenic Mustard cultivar developed in India by

(A) Delhi University (B) IARI-New Delhi

(C) Directorate of Oilseed, Hyderabad (D) BHU-Varanasi

96. The Bollgard Bt cotton cultivars was developed by

(A) Mahyco (B) Bayer

(C) Syngenta **(D) Monsanto**

97. The Bt cotton cultivars in India is supplied by

(A) Mahyco (B) Bayer

(C) Syngenta (D) None of the above

98. The Bt genes present in Brinjal

(A) Cry1Ac (B) Cry2Ab

(C) Cry1Ac and Cry2Ab (D) None of the above

99. The transgenic Bt cotton cultivars is

(A) Bikaneri Nerma (B) NHH-44

(C) Both A and B (D) None of the above

100. The transgenic crops approved for field trials in India are

(A) Rice (B) Brinjal

(C) Okra **(D) All of the above**

101. The Bt cotton was approved for commercial cultivation in India in the year

(A) 2002 (B) 2006

(C) 996 (D) 2004

102. The Bt Brinjal was approved for commercial cultivation in India in the year

(A) 2009 (B) 2011

(C) 2013 (D) 2015

103. The Bt Brinjal was banned for commercial cultivation in India in the year

(A) 2009 **(B) 2010**

(C) 2013 (D) 2015

104. Which was the first country to commercialize transgenic crops?

(A) USA (B) **China**

(C) Canada (D) Argentina

105. One of the Bt Cotton varieties approved for commercial cultivation under north Indian conditions is

(A) MECH 12 (B) MECH 162

(C) MECH 184 (D) **MECH 6304**

106. The Bt gene present in cotton cultivar Bollgard

(A) **Cry1Ac** (B) Cry2Ab

(C) Cry1Ac and Cry2Ab (D) None of the above

107. The Bt genes present in cotton cultivar Bollgard II

(A) Cry1Ac (B) Cry2Ab

(C) Cry1Ac and Cry2Ab (D) None of the above

108. The genes present in cotton cultivar Bollgard 3 are

(A) Cry1Ac (B) Cry2Ab

(C) Cry1Ac + Cry2Ab+Vip3A (D) None of the above

109. The Bt gene Cry1Ac is specific to

(A) **Lepidoptera** (B) Diptera

(C) Coleoptera (D) Hemiptera

110. The Bt gene Cry2Ab is specific to

(A) Lepidoptera and Hemiptera (B) **Diptera and Lepidoptera**

(C) Coleoptera and Diptera (D) Diptera and Hemiptera

111. The area under transgenic crops at the global level exceeded 100 million ha for the first time in

(A) 2004 (B) 2000

(C) **2006** (D) 2005

112. *Mayetiola destructor* is the scientific name of

(A) Grape phylloxera (B) Saw fly

(C) **Hessian fly** **(D)** Fruit fly

113. The largest area of a transgenic crop at global level is under

(A) Cotton (B) **Soybean**

(C) Maize (D) Rice

114. The largest area of a Bt transgenic crop at global level is under

(A) Cotton (B) Soybean

(C) **Maize** (D) Rice

115. The largest area under Bt cotton in India is in

(A) Andhra Pardesh (B) Punjab

(C) **Maharashtra** (D) Gujarat

116. Dual Discrimination Theory of host plat selection was proposed by

(A) G. S. Fraenkel (B) C. T. Bures

(C) M. Kogan (D) **G.G. Kennedy and J.D. Babour**

117. *Bacillus thruringiensis* was first isolated from silkworm in

(A) 1901 (B) 1912

(C) 1932 (D) 1942

118. Horizontal resistance is governed by

(A) Single genes (B) Few genes

(C) **Many Genes** (D) All the Above

119. Use of resistant varieties in the IPM is an example of

(A) Biological control (B) Chemical control

(C) **Cultural control** (D) Genetic control

120. The book "Host Plant Resistance to Insects" authored by

(A) N. Panda and G. S. Khush (B) G. S. Dhaliwal and U. K. Dilawari

(C) G. S. Dhaliwal and R. Singh (D) C. M. Smith

121. The first Book on HPR "Insect resistance in crop plants" authored by Painter in

(A) 1950 (B) 1953

(C) 1952 (D) **1951**

122. Insect plant interaction is based on .... trophic levels

(A) Four (B) **Three**

(C) Two (D) Five

123. Mechanisms of Resistance proposed by

(A) **Painter,1951** (B) Dethier,1953

(C) Thornstein,1960 (D) Fraenkel,1959

124. Plants which possess one or more additional genes are known as

(A) Protease inhibitors (B) Lectins

(C) Enzymes (D) **Transgenic Plants**

125. Gossypol cause antibiosis effect on

(A) *Ostrinia nubilalis* (B) *H. armigera* and *S. litura*

(C) *Ocimum basilicum* (D) ***H. viriscens* and *S. littoralis***

126. The author of book "Principles of Host Plant Resistance to Insects" is

(A) **N. Panda** (B) D. S. Dhaliwal

(B) R. Singh (D) C. M. Smith

127. The relative amount of heritable qualities possessed by a plant that reduces the degree of damage done to the plant by a pest or pests is

(A) Antibiosis (B) Antixenosis

(C) **Host plant resistance** (D) Tolerance

128. The type of resistance which is effective against all the known biotypes of the insects

(A) Specific Resistance (B) **Non Specific Resistance**

(C) Pure line resistance (D) Multiline resistance

129. The ability of the plant to with stand an insect population sufficient to damage severely the susceptible plants is called

(A) Antibiosis (B) Antixenosis

(C) **Host plant resistance** (D) Tolerance

130. A chemical released by an organism which is favourable to the emitter but not receiver is called

(A) **Allomone** (B) Synomone

(C) Pheromone (D) Kairomone

131. A chemical released by an organism which benefits both the sender and the receiver is called

(A) Allomone (B) **Synomone**

(B) Pheromone (D) Kairomone

132. Presence of which biochemical compound deter the larvae of European Corn borer from feeding and growth

(A) **DIMBOA** (B) Gossypol

(C) Salicylic (D) Sinigrin

133. Which of the following is a kairomone?

(A) Excitants (B) Attractants

(C) **Both (a) and (b)** (D) Repellents

134. Population capable of damaging and surviving on plants previously known to be resistant to other populations of the same species

(A) **Biotype** (B) Biospecies

(C) Biotic agents (D) None of the above

135. The phenomenon where a species is characterized by low reproductive rate and high survival rate is called

(A) **k–strategist** (B) r–strategist

(C) q–strategist (D) None of the above

136. Use of light traps for the management of pest is an example of

(A) **Mechanical control** (B) Chemical control

(C) Cultural control (D) Biological control

137. Use of resistant varieties constitutes method of

(A) **Genetic Control** (B) Natural control

(C) Biotic Control (D) Self Control

138. Vertical resistance is also called

(A) Oligogenic (B) Specific

(C) Monogenic (D) **All of the above**

139. Quarantine law was implemented in which year

(A) **1914** (B) 1916

(C) 1918 (D) 1971

140. Which of the following statement is not correct

(A) **All pests are pathogens**

(B) All Pathogens are pests

(C) Some parasites are beneficial to plants

(D) Some flowering plants are agricultural pests

141. The variety of tomato resistant to nematode is

(A) Pusa Ruby (B) Raj Kiran

(C) Azad (D) **S.l. 120**

142. Token stimuli theory was postulated by whom

(A) Throsteinson (B) Painter

(C) **Fraenkel** (D) None of the above

143. Vertical resistance is also called

(A) Oligogenic (B) Specific

(C) Monogenic (D) **All of the above**

144. Use of plant resistance in the IPM is an example of
(A) Biological control (B) Chemical Control
(C) **Cultural control** (D) Genetic control

145. Major gene resistance is also called
(A) Horizontal resistance (B) Oligogenic resistance
(C) **Vertical resistance** (D) Induced resistance

146. The type of resistance is effective against certain specific biotypes of the insect but not others
(A) Specific resistance (B) Vertical resistance
(C) **Both A and B** (D) Induced resistance

147. DIMBOA is considered to be resistance factor against European corn borer in
(A) **First generation** (B) Second generation
(C) Third generation (D) Fourth generation

148. A multiple insect and disease resistance variety which is grown on more than 11 million ha in the world is
(A) IR 6 (B) IR 16
(C) IR 26 (D) **IR 36**

149. Theory of chemosensory bases of host selection was proposedby
(A) **Dethier** (B) Hagstrum
(C) Morgan (D) Painter RH

150. Substance release by a non living material which is favourable to a receiving organism but determental to an organism of another species that is found in in or on the non living material is known as
(A) Pheromone (B) Kairomone
(C) **Aneumone** (D) Synomone

151. Use of light traps for the management of pest is an example
(A) Biological control (B) Chemical Control
(C) **Cultural control** (D) Mechanical control

152. The nutrition value of the plant and absence of toxic compound help in the process of
(A) Host finding (B) Host recognition
(C) Host acceptance (D) **Host suitability**

153. The correct sequence in host plant selection process in phytophagous insects is

(A) **Host habitat finding-Host finding-Host recognition-Host acceptance-Host suitability**

(B) Host habitat finding-Host finding-Host recognition-Host suitability-Host acceptance

(C) Host habitat finding-Host finding-Host acceptance-Host recognition-Host suitability

(D) None of the above.

154. Which of the following is a type of Ecological Resistance

(A) Host evasion (B) Induced resistance

(C) Escape (D) **All of the above**

155. A plant may pass through the most susceptible stage quickly or at time when insect number is less is called

(A) Host evasion (B) Induced resistance

(C) **Escape** (D) All of the above

156. Increase in resistance temporarily as result of some change conditions of plant or environment

(A) Host evasion (B) **Induced resistance**

(C) Escape (D) All of the above

157. The type of mechanism is which adverse effect of the host plant on the biology of the insects

(A) Non preference (B) Antixenosis

(C) **Antibiosis** (D) Tolerance

158. Resistance Horizontal resistance of governed by

(A) Single gene (B) Few Genes

(C) **Many genes** (D) All of the above

159. Resistance Which one of the following is a strategy to overcome development of biotypes

(A) Pyramiding of major genes

(B) Incorporation of horizontal resistance

(C) Breeding tolerant verities that exert no selection pressure on the insect.

(D) **All the above**

160. The number of biotypes in rice brown plant hopper is

(A) **5** (B) 6

(C) 7 (D) 8

161. The number of biotypes in rice gall midge is

(A) **13** (B) 10

(C) 8 (D) 12

162. Maximum number of insect biotypes of observed in

(A) Aphids (B) **Whiteflies**

(C) Midges (D) Hoppers

163. The occurrence of benzyl in minute quantities in barely was considered to be the primary causes of resistance to

(A) Aphids (B) **Green bugs**

(C) Midges (D) Borers

164. Which one of the following is an advantage of host plant resistance

(A) Specificity (B) Compatibility

(C) Decreases pesticide application (D) **All of the above**

165. Which one of the following is limitation of host plant resistance

(A) Time consuming (B) Biotype development

(C) Genetic limitations (D) **All of the above**

166. A polygenic trait enabling a plant to withstand or recover from arthropod damage without adversely affecting the growth or survival of the attacking arthropod

(A) Non preference (B) Antixenosis

(C) **Tolerance** (D) Antibiosis

167. Mayetiola destructor resistant wheat cultivar is

(A) PBW-343 (B) **Underhill**

(C) Winter Majetin (D) Allof the above

168. Winter majetin variety of apple was developed by

(A) JN Havens (B) RH Painter

(C) **Lindley** (D) Biffen

169. Underhill variety of wheat was developed by

(A) JN Havens (B) CV Riley

(C) Lindley (D) Biffen

170. Kawvale variety of wheat was developed by

(A) JN Havens (B) **RH Painter**

(C) Lindley (D) Biffen

171.Hessian fly resistant wheat variety is

(A) Underhill
(B) Kawvale
(C) **Both A and B**
(D) None of the above

172. The gene in common bean encoding arcelin, an amylase inhibitor inhibiting larval growth of *Zabrotes subfasciatus*

(A) Mir-gene
(B) **Arc-1 gene**
(C) Lep and AL genes
(D) All of the above

173. The gene in maize encoding cysteine proteinases that inhibits larval growth of *Spodoptera frugiperda*

(A) **Mir-gene**
(B) Arc-1 gene
(C) Lep and AL genes
(D) All of the above

174. The genes encoding Nonglycosylated steroidal alkaloids from foliage of wild potato, *Solanum chacoense*, that inhibit feeding of *Leptinotarsa decemlineata*

(A) Mir-gene
(B) Arc-1 gene
(C) **Lep and AL genes**
(D) All of the above

175. The gene controlling production of the C-glycosyl flavone maysin in maize silk that inhibits larval growth of *Helicoverpa zea*

(A) Mir-gene
(B) Arc-1 gene
(C) Lep and AL genes
(D) **pl -gene**

176. The DIMBOA which imparts chemical defense against the first generation of the European corn borer, *Ostrinia nubilalis* was first reported by

(A) JN Havens
(B) RH Painter
(C) Lindley
(D) **Klun and Brindley**

177. The expanded form of DIMBOA

(A) **2,4-dihydroxy-7- Ethoxy-2H-1,4- benzoxazin-3-one**
(B) 2,4-dihydroxy-7- methoxy-2H-1,4- benzoxazin-3-one
(C) 2,4-dimethoxy-7- methoxy-2H-1,4- benzoxazin-3-one
(D) 2,4-dimethoxy-7- ethoxy-2H-1,4- benzoxazin-3-one

178. The CC-NBS-LRR subfamily gene from tomato conferring resistance to *Meloidogyne incognita* and multiple insect species

(A) Mir-gene
(B) Arc-1 gene
(C) Lep and AL genes
(D) **Mi-1.2-gene**

179. The CC-NBS-LRR subfamily gene from melon conferring resistance to virus transmission by *Aphis gossypii*

(A) Mir-gene
(B) **Vat- gene**
(C) Lep and AL genes
(D) Mi-1.2-gene

180. Host Plant Seletion by phytophagousInscets Host Selection process it involves phases

(A) **5** (B) 6

(C) 7 (D) 8

181. Mechanism of host plant resistance involves

(A) Antixenosis (B) Antibiosis

(C) Tolerance **(D) All the above**

182. Types of biotypes Resistance are

(A) 2 (B) 4

(C) 6 (D) 3

183. The basis of antibiosis in corn against European corn borer is

(A) Nutrition imbalance

(B) **Presence of growth inhibitors MBOA and DIMBOA**

(C) Presence of toxins

(D) Nutrition deficiency

184. The basic of antibiosis in corn against *Heliothis zea* is

(A) Nutrition imbalance (B) Presence of growth inhibitor

(C) **Lethal silk factor** (D) Deficiency of nutrients

185. In leucerne resistance to spotted aphid is associated with

(A) Low sinigrin in leaves (B) High silica contents in leaves

(C) High MBOA contents in leaves (D) **High saponin in leaves**

186. Horizontal resistance is also known as

(A) Non-specific (B) General

(C) Polygenic (D) **All of the above**

187. The basis of antibiosis in corn against European corn borer is

(A) DIMBOA (B) MBOA

(C) Both A and B (D) All of the above

188. The Vertical resistance is also called

(A) Oligogenic (B) Specific

(C) Monogenic **(D) All of the above**

189. The token stimuli theory was postulated by

(A) Throsteinson (B) House

(C) Painter **(D) Fraenkel**

190. The biotype of BPH was first identified in Kerala on resistant veriety-
(A) IR-8 (B) IR-10
(B) IR-24 (D) **IR-26**

191. Which of *Entomophthora specics* is pathogenic to aphids
(A) *E.grylli* (B) ***E.coronata***
(C) *E.sphasperma* (D) None of the above

192. *Bacillus thuringiensis* is most effective at pH
(A) 3 to 6 (B) **9 to 10**
(C) 11 to 14 (D) None of the above

193. Which of the following order comprises phytophagous insects
(A) Odonata (B) Siphonoptera
(C) **Phasmida** (D) Mallophaga

194. The most important enzyme of MFO system is
(A) Phosphatase (B) **Cytochrome P-450**
(C) Carboxyesterase (D) Dechlorinase

195. The Insecticide having antagonistic effect with NPV
(A) DDT (B) Aldrin
(C) Endosulfan (D) **Malathion**

196. Glycogen is
(A) Monosaccharide (B) **Disaccharide**
(C) Oligosacchaide (D) Polysaccharide

197. Which of the following is a resistant variety of tomato against root knot nematode
(A) Pusa Ruby (B) **F455D1**
(C) Pusa Early (D) Punjab Kesari

198. Which of the folling enzyme is not associated with the digestion of food-
(A) Amylase (B) Cellulase
(C) Glucocidase (D) **Phenoloxidase**

199. The term antixenosis was proposed by
(A) **Smith** (B) Ortman
(C) Korgan and Ortman (D) None of the above

200. Dual discrimination theory was given by
(A) Painter (B) House
(C) **Kenedy and Booth** (D) Fraenkel

201. Which of the following is kairomone

(A) Excitants (B) Attractants

(C) **both A and B** (D) Repellents

202. The pest which occurs in isolated condition is

(A) **Sporadic pest** (B) Regular pest

(C) Key pest (D) Occasional pest

203. Aflatoxin is produced by

(A) *Bacillus* (B) ***Aspergillus***

(C) *Fusarium* (D) *Pythium*

204. Example of repellent used against termites

(A) Deet (B) Citronella oil

(C) **Pentachlorophenol** (D) Neem oil

205. In chemical beneficial to only releaser is called

(A) Kairomone (B) **Allomone**

(C) Synomone (D) Pheromone

206. In chemical beneficial to only receiver is called

(A) **Kairomone** (B) Allomone

(C) Synomone (D) Pheromone

207. A substance, which make the plant distasteful to insect feeding is called

(A) **Antifeedant** (B) Repellent

(C) Attractant (D) Pheromone

208. Father of host plant resistance

(A) **R.H.Painter** (B) R.H.Painter

(C) Paul Debach (D) Paul Muller

209. The pheromone secreted from which gland is the phenomena secreted

(A) Endocrine gland (B) **Exocrine gland**

(C) Both (D) None of these

210. Pheromone is also known as

(A) **Ectohormone** (B) Hormone

(C) Kairomone (D) None of these

211. Horizontal resistance is also known as

(A) Specific (B) **Non-specific**

(C) General (D) None of these

212.Insects feed on plants of several general within family known as

(A) Monophagous **(B) Oligophagous**

(C) Polyphagous (D) Phytophagous

213. All phytophagous insects are originally

(A) Monophagous (B) Oligophagous

(C) Polyphagous (D) Phytophagous

214. Pheromones are synthesized by

(A) Thoracic gland **(B) Glandular glands**

(C) Wax gland (D) Silk gland

215. The basis of antibiosis in corn against Helicoverpa zea

(A) Lethal silk factor (B) Nutritional imbalance

(C) Presence of growth inhibitor (D) None of these

216. The basis of antibiosis in maize European corn borer is

(A) Presence of growth inhibitors MBOA and DIMBOA

(B) Nutritional imbalance .

(C) Nutritional deficiency

(D) None of these

217. The number of biotypes in rice brown plant hopper is

(A) 5 (B) 6

(C) 7 (D) 8

218. The number of biotypes in rice gall midge is

(A) 04 (B) 05

(C) 06 (D) 08

219. Token stimuli theory was proposed by

(A) Painter (B) House

(C) Kenedy and Booth **(D) Fraenkel**

220. Vertical resistance is also called

(A) Specific (B) Non-specific

(C) Polygenic (D) None of these

221. The pest which is restricted to particular locality

(A) Sporadic pest (B) Regular pest

(C) Key pest **(D) Endemic pest**

222.Pest which occurs regularly and in serious manner is known as

(A) Sporadic pest (B) Regular pest

(C) **Epidemic pest** (D) Endemic pest

223. Which of the following act as Mites repellent

(A) Deet (B) Citronella oil

(C) Pentachlorophenol **(D) Benzyl benzoate**

224. Father of Host plant resistance is

(A) R.H.Painter (B) R.H. Whittaker

(C) Smith (D) None of these

225. The term Allelochemicals was coined by

(A) R.H.Painter **(B) R.H. Whittaker**

(C) Smith (D) None of these

226. Chemicals involved in communication are term as

(A) **Semiochemicals** (B) Allelochemicals

(C) Both A and B (C) None of these

227. The phenomenon in which species have high reproductive rate and low survival rate is called

(A) k-Strategist (B) **r-Strategist**

(C) u-Strategist (d) None of these

228. Which of the following is an advantage of host plant resistance

(A) Specificity (B) Compatibility

(C) None of these (D) **All of these**

229. Maximum number of insect biotypes is observed in

(A) BPH (B) Gall midge

(C) **Hessian fly** (D) Pea aphid

230. Maximum number of insect biotypes is observed in the insects group

(A) Aphids (B) Whiteflies

(C) Midges (D) Hoppers

231. The phenomenon in which species have low reproductive rate and high survival rate is called

(A) k-Strategist (B) r-Strategist

(C) Both A and B (D) None of these

232. An important major pest species in the complex of pests attacking a crop

(A) Sporadic pest (B) Regular pest

(C) **Key pest** (D) None of these

233. Yellow sticky trip is used for attracting

(A) **Aphids** (B) Leaf miners

(C) Scale insects (D) Mealy bugs

234. The earliest documented case of insect resistance is against

(A) Wheat stem sawfly (B) **Hessian fly**

(C) Sorghum shoot fly (D) Screw worm fly

235. The non-preference plant resistance to insect is also known as

(A) Antibiosis (B) **Antixenosis**

(C) Tolerance (D) Host evasion

236. Use of resistance varieties in the IPM is an example of

(A) Biological control (B) Chemical control

(C) **Cultural control** (D) Genetic control

237. Major gene resistance is also called

(A) Horizontal resistance (B) Oligogenic resistance

(C) **Vertical resistance** (D) Induced resistance

238. DIMBOA is considered to be resistance factor against European corn borer is

(A) **First generation** (B) Second generation

(C) Third generation (D) Fourth generation

239. Theory of chemosensory bases of host selection was proposed by

(A) Dethier (B) Morgan

(C) Hagstrum (D) none of these

240. In which of the following host response are involved

(A) Antibiosis (B) Antixenosis

(B) **Tolerance** (D) all of the above

241. Resistance develop in the host plants due to natural enemies in known as

(A) Intrinsic resistance (B) **Extrinsic resistance**

(C) Both A and B (D) None of these

242. Ecological resistance is also known as

(A) Genetic resistance (B) **Pseudo resistance**

(C) Evolutionary resistance (D) All of these

243. Botanical instinct theory was given by

(A) House (B) Painter
(C) **Brues** (D) Jeremy

244. Endotoxin is produces by

(A) Gram negative bacteria Bt **(B) Gram positive bacteria Bt**
(C) Both A and B (D) None of these

245. Which of the following is involved in Ecological resistance

(A) Host evasion (B) Induced resistance
(C) Host escape (D) **All of these**

246. In chemical beneficial to both releaser and receiver is callad

(A) Kairomone (B) Allomone
(C) **Synomone** (D) Pheromone

247. The gene Barstar, Bar, Barnase are used for developing form GM

(A) Soyabean (B) Cotton
(C) Maize (D) **Mustard**

Unit-7

Biological Control

Q 1. Match the following

1. The book *"Biological Control of Insect Pests and Weeds"* is authored by

(A) CB Huffaker and PS Messenger (B) **Paul de Bach**

(C) HD Burges and NW Hussey (D) TS Bellows Jr

2. The book *"Biological Control"* is authored by

(A) **Van Drische and TS Bellows Jr** (B) Paul de Bach

(C) HD Burges and NW Hussey (D) Huffaker and Messenger

3. The book *"Biological Control of Insect and Mites"* is authored by

(A) CB Huffaker and PS Messenger (B) Paul de Bach

(C) **HD Burges and NW Hussey** (D) TS Bellows Jr

4. The book *"Theory and Practice of Biological Control"* is authored by

(A) **CB Huffaker and PS Messenger** (B) Paul de Bach

(C) HD Burges and NW Hussey (D) TS Bellows Jr

5. The author of *"Principles of Insect Pathology"* is

(A) HC Coppel and WM James (B) **E. A. Steinhaus**

(C) HD Burges and NW Hussey (D) TS Bellows Jr

6. The author of *"Microbial Control of Insects and Mites"* is

(A) HC Coppel and WM James (B) E. A. Steinhaus

(C) **HD Burges and NW Hussey** (D) TS Bellows Jr

7. The author of *"Biological Insect Pest Suppression"* is

(A) **HC Coppel and WM James** (B) E. A. Steinhaus

(C) HD Burges and NW Hussey (D) TS Bellows Jr

8. The term Biological control was used by

(A) CB Huffaker (B) Paul de Bach

(C) **A. Harry Scott Smith** (D) TS Bellows Jr

9. The principal areas of applicability of biocontrol are
   (A) Where the value of crop is too low
   (B) Where overuse of insecticide necessitates use of other methods
   (C) Where biocontrol is not a part a part of already IPM programme
   (D) **All of the above**

10. The objectives of Biological control is
   (A) Reduction of pest population below EIL
   (B) Prevention of pest build up
   (C) Delay in pest build up
   (D) All of the above

11. The success rate of natural enemies in pest suppression worldwide is about
   (A) **38%** (B) 48%
   (C) 58% (D) 28%

12. Commonwealth Institute for Biological Control (CIBC) is located at
   (A) UK (B) USA
   (C) **Trinidad** (D) Belgium

13. The international organization for Biological Control (CIBC) is
   (A) **IOBC** (B) CSIRO
   (C) CIBC (D) ANBCP

14. The organization for Biological Control in India is
   (A) **NBAIR** (B) ICAR
   (C) NCIPM (D) DPPQ&S

15. The example of successful complete control using biocontrol agents
   (A) **Cottony Cushion scale using Vedalia beetle**
   (B) *Aphis craccivora* using *Menochilus sexmaculatus*
   (C) Bollworm complex using *Trichogramma* sp.
   (D) All of the above

16. The example of successful substantial control using biocontrol agents
   (A) Cottony Cushion scale using Vedalia beetle
   (B) ***Aphis craccivora* using *Menochilus sexmaculatus***
   (C) Bollworm complex using *Trichogramma* sp.
   (D) All of the above

17. The example where partial control using biocontrol agents is achieved
    (A) Cottony Cushion scale using Vedalia beetle
    (B) *Aphis craccivora* using *Menochilus sexmaculatus*
    (C) **Bollworm complex using *Trichogramma* sp.**
    (D) All of the above

18. The example of extinction of species due to biocontrol agents
    (A) *Lerrvana iridiscens*
    (B) Coconut moth
    (C) **Both A and B**
    (D) None of the above

19. The extinction of *Lerrvana iridiscens*, coconut moth due to Tachinid moth, *Bessa remota* was reported in
    (A) Japan (1931)
    (B) **Fiji (1925)**
    (C) Malaysia (1925)
    (D) None of the above

20. The maintenance of a more or less fluctuating population density of an organism within certain definable upper and lower limits over a period of time by actions of abiotic and/ or biotic environmental factors is termed as
    (A) Classical Control (B) Biological Control
    (C) Fortitudinous Control (D) **Natural control**

21. Importation of potential natural enemies for the purpose of control of a particular exotic or indigenous pest which has established well in an area is called
    (A) Classical Control (B) Biological Control
    (C) **Fortitudinous Control** (D) Natural Control

22. The utilization of organisms for the control of population densities of animals and plants
    (A) Classical Control (B) **Biological Control**
    (C) Fortitudinous Control (D) Natural Control

23. The Indian Mynah bird, *Gracula religiosa* was exported from India to Mauritius to control
    (A) Red locust (B) Mealy bug
    (C) **Coffee green scale** (D) Cottony cushion scale

24. The Indian Mynah bird was exported from India to Mauritius to control Red locust in the year

(A) **1762** (B) 1788
(C) 1888 (D) 1929

25. The Vedalia beetle, *Rodolia cardinalis* was exported from Australia to California (USA) in the year

(A) 1762 (B) 1788
(C) **1888** (D) 1929

26. The Vedalia beetle, *Rodolia cardinalis* was exported from Australia to California (USA) to control

(A) *Aphelinus mali* (B) ***Icerya purchasi***
(C) *Malacosoma indicum* (D) None of the above

27. The Indian Mynah bird, *Gracula religiosa* was exported from India to Mauritius to control

(A) Red locust (B) Mealy bug
(C) **Coffee green scale** (D) Cottony cushion scale

28. The Indian Mynah bird, *Gracula religiosa* was exported from India to Mauritius to control coffee green scale in the year

(A) 1888 (B) **1762**
(C) 1909 (D) 1862

29. The lady birdbeetle, *Cryptolaemus montrouzieri* was exported from Australia to India to control

(A) Cottony cushion scale (B) **Coffee green scale**
(C) Coconut black headed caterpillar (D) Wooly apple aphid

30. The lady birdbeetle, *Cryptolaemus montrouzieri* was imported from Australia to India to control coffee green scale in the year

(A) 1888 (B) 1762
(C) 1909 (D) **1898**

31. The *Aphelinus mali* was imported from North America to India (Tamil Nadu) to control

(A) *Icerya purchasi* (B) *Coccus viridis*
(C) *Opisina arenosella* (D) ***Eriosoma lanigerum***

32. The *Aphelinus mali* was imported from North America to India (Tamil Nadu) to control WAA (*Eriosoma lanigerum*) in the year

(A) 1888 (B) 1762
(C) 1909 (D) **1937**

33. The Tachinid, *Bessa remota* was imported from Srilanka to India to control

(A) *Icerya purchasi* (B) *Coccus viridis*

(C) ***Opisina arenosella*** (D) *Eriosoma lanigerum*

34. The Tachinid, *Bessa remota* was imported from Srilanka to India to control Coconut black headed caterpillar, *Opisina arenosella* in the year

(A) 1888 (B) 1762

(C) **1960** (D) 1937

35. The earliest known fossil of a parasitoid was recorded in

(A) **150 million years ago** (B) 100 million years ago

(C) 250 million years ago (D) 350 million years ago

36. The earliest known fossil of predator was recorded in

(A) 150 million years ago (B) 100 million years ago

(C) **280 million years ago** (D) 350 million years ago

37. The term parasitoid was coined by

(A) Smith (B) Paul De Bach

(C) Huffaker (D) **Reuter**

38. Among the natural enemies, which group is the largest

(A) **Parasitiods** (B) Predators

(C) Parasites (D) Pathogens

39. The insect order having largest number of parasitoids is

(A) **Hymenoptera** (B) Diptera

(C) Coleoptera (D) Neuroptera

40. The insect order being parasitoid of Thysanura

(A) Hymenoptera (B) **Strepsiptera**

(C) Coleoptera (D) Neuroptera

41. The important characteristic of Parasitoid is

(A) Parasitic behaviour is expressed only in the larval stage of the parasitoid

(B) The adult stage is free living

(C) A parasitoid usually destroys its host as it develops to maturity

(D) **All of the above**

42. The desirable attributes of Parasitoid is

(A) High searching ability

(B) Host specificity

(C) High fecundity and short development time

(D) **All of the above**

43. The parasitoid develop singly per host is

(A) Solitary endoparasitoid (B) **Solitary parasitoid**

(C) Gregarious parasitoid (D) Gregarious parasitoid

44. When more than one or several parasitoid progeny of a single species habitually develop per host

(A) Solitary endoparasitoid (B) Solitary parasitoid

(C) Gregarious endoparasitoid (D) **Gregarious parasitoid**

45. When more than one or several parasitoid progeny of a single species habitually develop per host

(A) Solitary endoparasitoid (B) Solitary parasitoid

(C) Gregarious parasitoid (D) **Gregarious parasitoid**

46. Parasitoids are specific to a host

(A) **Monophagous parasitoid** (B) Oligophagous parasitoid

(C) Polyphagous parasitoid (D) None of the above

47. Example of Monophagous Parasitoids

(A) ***Platygaster*** (B) *Telenomus*

(C) *Bracon* (D) None of the above

48. Parasitoids having host range is restricted within a genus of the host species

(A) Monophagous parasitoid (B) **Oligophagous parasitoid**

(C) Polyphagous parasitoid (D)` None of the above

49. Example of Oligophagous Parasitoids

(A) *Platygaster* (B) ***Telenomus***

(C) *Bracon* (D) None of the above

50. Parasitoids having broad host range which may comprise few to many genera.

(A) Monophagous parasitoid (B) Oligophagous parasitoid

(C) **Polyphagous parasitoid** (D) None of the above

51. Example of Polyphagous Parasitoids

(A) *Platygaster* (B) *Telenomus*

(C) ***Bracon*** (D) None of the above

52. Example of egg parasitoid is

(A) ***Trichogramma* sp.** (B) *Epricania melanoleuca*

(C) *Isotima javensis* (D) *Brachymeria nephantidis*

53. *Trichogramma chilonis*, an egg parasitoid is used for the control of

(A) Sugarcane internode borer (B) Rice leaf folder

(C) Bollworm in cotton (D) **All of the above**

54. *Trichogramma embryophagum*, an egg parasitoid is used for the control of

(A) Sugarcane internode borer (B) Rice leaf folder

(C) Bollworm in cotton (D) **Apple coddling moth**

55. *Trichogramma brasiliensis*, an egg parasitoid is used for the control of

(A) Sugarcane internode borer (B) Rice leaf folder

(C) Bollworm in cotton (D) **Tomato fruit borer**

56. *Trichogramma acheae*, an egg parasitoid is used for the control of

(A) Sugarcane internode borer (B) Rice leaf folder

(C) **Spotted Bollworm** (D) None of the above

57. *Trichogramma japonicum*, an egg parasitoid is used for the control of

(A) Sugarcane internode borer (B) Rice leaf folder

(C) Bollworm in cotton (D) **Rice stem borer**

58. The size of Trichocard is

(A) 5x10cm (B) **10x15cm**

(C) 18x9cm (D) 9x18cm

59. The One 1 cc of Trichocard contains

(A) 10000-15000 pa rasitized eggs

(B) **18000-20000 parasitized eggs**

(C) 20000-25000 parasitized eggs

(D) 25000-28000 parasitized eggs

60. The number of Trichocard required per hectare is

(A) **15-25** (B) 25-40

(C) 40-60 (D) 60-75

61. *Telenomus remus*, an egg parasitoid is used for the control of

(A) Sugarcane internode borer (B) Rice leaf folder

(C) ***Sopodoptera litura*** (D) Rice stem borer

62. Example of Larval parasitoid is

(A) *Trichogramma* sp. (B) *Epricania melanoleuca*

(C) ***Campoletis chloridae*** (D) *Brachymeria nephantidis*

63. *Campoletis chloridae,* a larval parasitoid is used for the control of

(A) Sugarcane internode borer (B) Spotted boll worm

(C) *Sopodoptera litura* (D) ***Helicoverpa armigera***

64. *Platygaster oryzae,* a larval parasitoid is used for the control of

(A) Sugarcane internode borer (B) Rice leaf folder

(C) **Rice gall midge** (D) Rice stem borer

65. *Cyrtorhinus lividipennis,* is used for the control of

(A) **Brown Plant Hopper** (B) Rice leaf folder

(C) ice gall midge (D) Rice stem borer

66. *Cyrtorhinus lividipennis,* is used for the control of

(A) **Brown Plant Hopper** (B) Rice leaf folder

(C) Rice gall midge (D) Rice stem borer

67. Example of Adult parasitoid is

(A) *Trichogramma* sp.. (B) ***Epricania melanoleuca***

(C) *Isotima javensis* (D) *Brachymeria nephantidis*

68. *Epricania melanoleuca*, an adult parasitoid is used for the control of

(A) Sugarcane internode borer (B) Rice leaf folder

(C) **Sugarcane pyrilla** (D) Rice stem borer

69. Example of Pupal parasitoid is

(A) *Trichogramma* sp. (B) *Epricania melanoleuca*

(C) *Isotima javensis* (D) ***Brachymeria nephantidis***

70. *Brachymeria nephantidis*, a pupal parasitoid is used for the control of

(A) Sugarcane internode borer (B) Rice leaf folder

(C) **Coconut black headed caterpillar** (D) Rice stem borer

71. Example of Egg-Larval parasitoid is

(A) ***Chelonus blackburnii*** (B) *Isotima javensis*

(C) *Aphelinus mali* (D) *Encarisa formosa*

72. *Chelonus blackburnii*, an egg-larval parasitoid is used for the control of

(A) Sugarcane internode borer (B) Rice leaf folder

(C) **Spotted bollworm** (D) Rice stem borer

73. The release rate of *Chelonus blackburnii,* for PTM control is

(A) Two release in the field (B) 5 adults/5 kg potatoes in godowns @50,000/release

(C) **Both A and B** (D) None of the above

74. Example of Larval-pupal parasitoid is
   (A) *Chelonus blackburnii* (B) ***Isotima javensis***
   (C) *Aphelinus mali* (D) *Encarisa formosa*

75. *Isotima javensis*, a larval-pupal parasitoid is used for the control of
   (A) Sugarcane internode borer (B) Rice leaf folder
   (C) **Sugarcane top borer** (D) Rice stem borer

76. Example of Pupal-Adult parasitoid is
   (A) *Chelonus blackburnii* (B) *Isotima javensis*
   (C) ***Aphelinus mali*** (D) *Encarisa formosa*

77. *Aphelinus mali*, a pupal-adult parasitoid is used for the control of
   (A) Coddling moth (B) **Wooly apple aphid**
   (C) San Jose scale (D) Rice stem borer

78. Example of Nymphal-Adult parasitoid is
   (A) *Chelonus blackburnii* (B) *Isotima javensis*
   (C) *Aphelinus mali* (D) ***Encarisa formosa***

79. *Encarsia formosa*, a Nymphal-adult parasitoid is used for the control of
   (A) *Spodoptera litura* (B) *Aphis craccivora*
   (C) *Helicoverpa armigera* (D) ***Bemesia tabaci***

80. The parasitoids which lead to arrest or termination of the host development immediately after attack known as
   (A) **Idiobionts** (B) Koinobionts
   (C) Synobionts (D) none of the above

81. The parasitoids which permanently impair or paralyze the hosts after attack and oviposition and thereby donot prevent their further development are known as
   (A) Idiobionts (B) **Koinobionts**
   (C) Synobionts (D) none of the above

82. The condition resulting out of the attack of a single host by a primary parasitoid
   (A) **Primary parasitism** (B) Multiple parasitism
   (C) Super parasitism (D) Clepto parasitism

83. The condition resulting from the simulataneous use of a single host individual by two or more species of primary parasitoids
   (A) Primary parasitism (B) **Multipleparasitism**
   (C) Super parasitism (D) Clepto parasitism

84. The condition resulting from the simulataneous use of a single host individual by parasitoids of the same species

(A) Primary parasitism (B) Multiple parasitism

(C) **Super parasitism** (D) Clepto parasitism

85. The condition in which the adult parasitoid preferentially appropriates itself for its own progeny, the previously paralysed and parasitized host of another parasitoid

(A) Primary parasitism (B) Multiple parasitism

(C) Super parasitism (D) **Clepto parasitism**

86. The condition resulting from parasitoids using their own species as a host to develop some of their progeny

(A) Primary parasitism (B) Multiple parasitism

(C) **Adelpho parasitism** (D) Clepto parasitism

87. The condition where an organism that is a parasite of a parasite

(A) Primary parasitism (B) **Hyperparasitism**

(C) Adelpho parasitism (D) Clepto parasitism

88. The important characteristic of Predators is

(A) Both adult and young stage are free living

(B) The eggs of the predators are usually laid in the vicinity of the host

(C) The will consume a number of prey for completing its life cycle.

(D) **All of the above**

89. In general, predators are carnivorous in both adult and young stage except

(A) *Cyrtorhinus lividipennis* (B) *Chrysoperla carnea*

(C) **Both A and B** (D) none of the above

90. Predators feeding on only one category of prey

(A) **Monophagous** (B) Oligophagous

(C) Polyphagous (D) None of the above

91. Example of Monophagous Predator

(A) ***Dipha aphidovora*** (B) *Episyrphus balteatus*

(C) *Chrysoperla carnea* (D) None of the above

92. Predators feeding on a narrow range of prey spread within a genus

(A) Monophagous (B) **Oligophagous**

(C) Polyphagous (D) None of the above

93. Example of Oligophagous Predator

(A) *Dipha aphidovora* (B) ***Episyrphus balteatus***

(C) *Chrysoperla carnea* (D) None of the above

94. Predators feeding a range of prey and are generalists in nature

(A) Monophagous (B) Oligophagous

(C) **Polyphagous** (D) None of the above

95. Example of Polyphagous Predator

(A) *Dipha aphidovora* (B) *Episyrphus balteatus*

(C) ***Chrysoperla carnea*** (D) None of the above

96. Defense mechanism in insects against natural enemies are

(A) Encapsulation (B) Melanization

(C) Humeral reaction (D) **All of the above**

97.Biological control concerned with introduction of natural enemies

(A) **Classical biological control** (B) Augmentative biological control

(C) Conservation biological control (D) All of the above

98. Biological control practiced by controlled releases of natural enemies

(A) Classical biological control (B) **Augmentative biological control**

(C) Conservation biological control (D) All of the above

99. Biological control concerned with the avoiding practices detrimental to the natural enemies performance

(A) Classical biological control (B) Augmentative biological control

(C) **Conservation biological control** (D) All of the above

100. The order of relative importance type of biological control is

(A) **Conservation > Augmentation > Classical approach**

(B) Classical approach > Augmentation > Conservation

(C) Conservation > Classical approach > Augmentation

(D) Augmentation > Classical approach > Conservation

101. Biological control practiced by releases of exotic natural enemies against exotic pest is

(A) **Classical biological control** (B) Augmentative biological control

(C) Conservation biological control (D) All of the above

102. Biological control practiced by releases of exotic natural enemies against native pest is
(A) **Classical biological control** (B) Neoclassical biological control
(C) Conservation biological control (D) All of the above

103.Biological control practiced by releases of exotic natural enemies usually not associated with the against exotic pest is
(A) Classical biological control
(B) Neoclassical biological control
(C) **New Association Classical biological control**
(D) All of the above

104. The concept of augmentative release was given by
(A) **De Bach and Hagen** (B) Hale and Elliot
(C) Huffaker (D) None of the above

105. Augmentative release is more successful in
(A) **Green house** (B) Open field
(C) Both Open and Green house (D) None of the above

106. Characteristics of agents for augmentation
(A) Adaptation to environment (B) High searching ability
(C) Synchrony with host/prey cycle (D) **All of the above**

107. The use of living organisms to control pests when control is achieved exclusively by the released organisms themselves
(A) Inoculative release method (B) **Inundative release method**
(C) Supplementary release (D) All of the above

108. The use of living organisms to control pests when control is achieved by the subsequent progenies of the released organisms
(A) **Inoculative release method** (B) Inundative release method
(C) Supplementary release (D) All of the above

109. NEs Release are made when sampling indicates that a pest is about to escape control by its NEs
(A) Inoculative release method (B) Inundative release method
(C) **Supplementary release** (D) All of the above

110. The Inoculative release method is more successful where
(A) Crop is of longer duration (B) Environment is stable
(C) **Both A and B** (D) All of the above

111. The Inoculative release method is more successful where
(A) Crop is of shorter duration (B) Environment is less stable
(C) **Both A and B** (D) All of the above

112. The conservation of natural enemies can be achieved by
(A) Rationalizing pesticide use (B) Provision of food and shelter
(C) Beneficial management practices (D) **All of the above**

113. Successful biological control programmes report benefit: cost ratios from reduced pest impacts and decreased use of pesticides ranges
(A) **10-30: 1** (B) 30-50 : 1
(C) 50-70:1 (D) None of the above

114. The largest and most important Parasitoid family of Diptera
(A) **Tachinidae** (B) Syrphidae
(C) Asilidae (D) Sarcophagidae

115. The Ichneumonid parasitoids are characterized by
(A) Long Antennae>13 segmented
(B) Trochantellus present
(C) Ovipositor long often longer than body
(D) **All of the above**

116. The Chalcidid parasitoids are characterized by
(A) Hind femur stout and swollen with rows of short teeth
(B) wings not folded longitudinally at rest
(C) Ovipositor short and straight
(D) **All of the above**

117. The Trichogrammatid parasitoids are characterized by
(A) 3 segmented tarsi
(B) Antennae 5-9 segmented
(C) Forewings lack a post marginal vein ; setae are arranged in radiating l ine
(D) **All of the above**

118. The Eulophid parasitoids are characterized by
(A) 4 Segmented tarsi
(B) Non pedunculate wings
(C) Forewing with marginal vein long; post marginal and stigmal veins short
(D) **All of the above**

119. The *Pediobus foveolatus* is a parasitoid of

(A) ***Henosepilachna vigintioctpunctata*** (B) *Helicoverpa armigera*

(C) *Opsinia arseonella* (D) San Jose scale

120. The Mantids are characterized by

(A) Raptorial forelegs (B) Pronotum not covered the head

(C) Proventriculus is not armed with teeth (D) **All of the above**

121. The Reduviids are characterized by

(A) Rostrum 3- segmented (B) Forelegs bear adhesive pads on tibia

(C) Metathoracic scent glands absent (D) **All of the above**

122. The Coccinellids are characterized by

(A) Tarsi 4 segmented and 3rd segment concealed in 2nd bilobed segment
(B) Head concealed in pronotum
(C) Larvae campodeiform covered with minute tubercles or spines
(D) **All of the above**

123. The Chrysopids are characterized by

(A) Bright green body with green wings and iridiscent eyes

(B) Antennae filiform

(C) Egg are stalked

(D) **All of the above**

124. The *Phytoseiulus persimilis*is a predator of

(A) ***Mites*** (B) Aphids

(C) Jassids (D) Scale

125. To control bean mite, *Phytoseiulus persimilis* is release at the rate of

(A) **10 adults/plant** (B) 20 adults/plant

(C) 50 adults/plant (D) None of the above

126. Parasitiods provides effective biocontrol than predators because

(A) Only one or fewer host is required for its complete development
(B) Population can be sustained at low levels
(C) Short life cycle and synchronized with that of pest density
(D) **All of the above**

127. The term Neocalssical biological control was coined by

(A) **Lockwood** (B) Paul De Bach

(C) Bellows and Fisher (D) None of the above

128. For controlling rice stem borer, *Trichogramma japonicum* is released at the rate of

(A) **50,000-1,00,000/ha** (B) 1,00,000-1,50,000/ha
(C) 30,000-50,000/ha (D) None of the above

129. For controlling *Helicoveerpa armigera*, *Trichogramma chilonis* is released at the rate of

(A) 50,000-1,00,000/ha (B) 1,00,000-1,50,000/ha
(C) **1,50,000/ha** (D) None of the above

130. For controlling Sugarcane Pyrilla, *Epiricania melanoleuca* is released at the rate of

(A) 2-3 egg mass/40 selected spots (B) 5-7 cocoons/40 selected spots
(C) **Both A and B** (D) None of the above

131. For controlling *Aphis craccivora*, *Chrysoperla carnea* is released at the rate of

(A) **2nd instar grub @ 2000/acre** (B) 3rd instar grub @ 5000/acre
(C) Both A and B (D) None of the above

132. For controlling *Spodoptera litura*, *Telenomus remesus* is released at the rate of

(A) **1.20 Lakh/ha** (B) 1.50 lakh/ha
(C) 2.00 lakh/ha (D) None of the above

133. For controlling Gall midge, *Platygaster oryzae* is released at the rate of

(A) **1 parasitized gall/10m²** (B) 5 parasitized gall/10m²
(C) 10 parasitized gall/10m² (D) None of the above

134. For controlling Sugarcane early shoot borer, *Strumpiopsis inferens* is released at the rate of

(A) **125 gravid females/ha** (B) 500gravid females/ha
(C) 250 gravid females/ha (D) None of the above

135. For controlling San Jose Scale, *Encarsia perniciosi* is released at the rate of

(A) 500 adults/infested tree (B) 1000 adults/infested tree
(C) 1500 adults/infested tree (D) **2000 adults/infested tree**

136. For controlling Coddling Moth, *Trichogramma embryophagum* is released at the rate of

(A) 500 adults/infested tree (B) 1000 adults/infested tree
(C) 1500 adults/infested tree (D) **2000 adults/infested tree**

137. For controlling Citrus mealybug, *Cryptolaemus montrouzieri* is released at the rate of

(A) 5 beetles/Plant (B) **10 beetles/Plant**

(C) 15 beetles/Plant (D) None of the above

138. For controlling Grape mealy bug, *Cryptolaemus montrouzieri* is released at the rate of

(A) 5 beetles/vine (B) **10 beetles/vine**

(C) 15 beetles/vine (D) None of the above

139. *Copidosoma koehleri*, an egg-larval parasitoid is used for the control of

(A) Sugarcane internode borer (B) Rice leaf folder

(C) Bollworm in cotton (D) **Potato tube moth**

140. *Aphytis diaspidis* parasitoid is used for the control of

(A) **San Jose Scale** (B) Cottony cushion scale

(C) Wooly apple aphid (D) Planococcus citri

141. *Platymeris laevicollis* is used for the control of

(A) San Jose Scale (B) Cottony cushion scale

(C) **Coconut Rhinoceros beetle** (D) *Planococcus citri*

142. *Leptomastix dactylopii* is used for the control of

(A) *Achea janata* (B) *Coccus viridis*

(C) *Hypothenemus hampei* (D) ***Planococcus citri***

143. *Telenomus remesus* is used for the control of

(A) ***Achea janata*** (B) *Coccus viridis*

(C) *Hypothenemus hampei* (D) *Planococcus citri*

144. *Phymastichus coffea* is used for the control of

(A) *Achea janata* (B) *Aleurodicus dispersus*

(C) ***Hypothenemus hampei*** (D) *Planococcus citri*

145. *Bracon kirkpatricki* is used for the control of

(A) ***Pectinophora gossypiella*** (B) *Helicoverpa armigera*

(C) *Erias vitella* (D) *Erias insulana*

146. *Cotesia flavipes* is a parasitoid of

(A) *Chilo sacchariphagous indicus* (B) *Helicoverpa armigera*

(C) ***Chilo auricilius*** (D) *Scirpophaga incertulas*

147. *Diatraeophaga striatalis* is a parasitoid of

(A) ***Chilo sacchariphagous indicus*** (B) *Helicoverpa armigera*

(C) *Chilo auricilius* (D) *Scirpophaga incertulas*

148. *Isotima javensis* is a parasitoid of

(A) *Chilo sacchariphagous indicus* (B) *Helicoverpa armigera*

(C) *Chilo auricilius* (D) ***Scirpophaga excerptalis***

149. *Acerophagus papayae* is a parasitoid of

(A) ***Paracoccus marginatus*** (B) *Helicoverpa armigera*

(C) *Chilo auricilius* (D) *Scirpophaga excerptalis*

150. *Trichogramma* sp. is reared on

(A) ***Corcyra cephalonica*** (B) *Spodoptera litura*

(C) *Phthorimaea operculella* (D) Mealybugs

151. *Cryptolaemus montrouzieri* is reared on

(A) *Corcyra cephalonica* (B) *Spodoptera litura*

(C) *Phthorimaea operiulella* (D) **Mealybugs**

152. *Copidosoma koehleri* sp. is reared on

(A) *Corcyra cephalonica* (B) *Spodoptera litura*

(C) ***Phthorimaea operculella*** (D) Mealybugs

153. *Goniozus nephantidis* is reared on

(A) *Corcyra cephalonica* (B) **Coconut black head caterpillar**

(C) *Phthorimaea operiulella* (D) Mealybugs

154. *Telenomus remesus* is reared on

(A) *Corcyra cephalonica* (B) ***Spodoptera litura***

(C) *Phthorimaea operiulella* (D) Mealybugs

155. *Chrysoperla carnea* is reared on

(A) ***Corcyra cephalonica*** (B) *Spodoptera litura*

(C) *Phthorimaea operiulella* (D) Mealybugs

156. *Bacillus thuringiensis* was first isolated by

(A) Berliner (B) **Ishiwata**

(C) Ignoffo (D) None of these

157. *Bacillus thuringiensis* was named as *Bacillus thuringiensis* by

(A) **Berliner** (B) Ishiwata

(C) Ignoffo (D) None of these

158. *Bacillus thuringiensis* was first isolated by S. Ishiwata in the year 1905 from

(A) *Ephestia kuhniella* (B) ***Bombyx mori***
(C) *Galleria melonella* (D) None of these

159. *Bacillus thuringiensis* is a

(A) Gram *–ve* (B) **Gram +*ve***
(C) Both A and B (D) None of these

160. *Bacillus thuringiensis* is a

(A) **Stomach poison** (B) Contact poison
(C) Systemic poison (D) None of these

161. The most successful isolate of *Bacillus thuringiensis* var *kurstaki* is

(A) **HD 1** (B) HD 2
(C) HD 3 (D) HD 4

162. *thuringiensis* subsp. *kurstaki* is pathogenic to

(A) **Lepidoptera** (B) Coleoptera
(C) Diptera (D) None of these

163. *Bacillus thuringiensis* subsp. *galleriae* is pathogenic to

(A) Scarabaeids (B) **Waxmoth**
(C) Beetle and weevils (D) Simulids

164. *Bacillus thuringiensis* subsp. *israelensis*. is pathogenic to

(A) Lepidoptera (B) Coleoptera
(C) **Diptera** (D) None of these

165. *Bacillus thuringiensis* subsp. *morrisoni* is pathogenic to

(A) Lepidoptera (B) **Coleoptera**
(C) Diptera (D) None of these

166. The Cry I toxin is pathogenic to

(A) **Lepidoptera** (B) Coleoptera
(C) Diptera (D) None of these

167. The classification of crystal protein was given by

(A) Feitelson (B) **Hofte and Whiteley**
(C) Kreig *et al.* (D) None of these

168. The Cry II toxin is pathogenic to

(A) **Lepidoptera and Diptera** (B) Coleoptera
(C) Diptera (D) None of these

169. The Cry III toxin is pathogenic to

(A) Lepidoptera (B) **Oleoptera**

(C) Diptera (D) None of these

170. The Cry IV toxin is pathogenic to

(A) Lepidoptera (B) Coleoptera

(C) **Diptera** (D) None of these

171. The Cry V toxin is pathogenic to

(A) Lepidoptera (B) **Lepidoptera and Coleoptera**

(C) Diptera (D) None of these

172. The Cry VI toxin is pathogenic to

(A) Lepidoptera (B) Coleoptera

(C) Diptera (D) **Nematodes**

173. Which of the following is obligate spore former

(A) ***Bacillus popillae*** (B) *Bacillus thuringiensis*

(C) *Bacillus ceres* (D) None of these

174. The Milky disease is caused by

(A) ***Bacillus popillae*** (B) *Bacillus thuringiensis*

(C) *Bacillus ceres* (D) None of these

175. The Milky disease was discovered by Dutky in 1937 in

(A) Japanese beetle (B) *Popillia japonica*

(C) **Both A and B** (D) None of these

176. *Bacillus popillae* is used to control

(A) *Holotrichia consanguinea* (B) *Leucopholis coneophora*

(C) *Leucophora lepidophora* (D) **All of the above**

177. The first of formulation of *Bt* was sold under the trade name of

(A) Bactericide (B) **Thuricide**

(C) Btcide (D) None of these

178. The total area under Bt cotton on India is

(A) 9.60 mha (B) **11.60 mha**

(C) 14.60mha (D) 17.60 mha

179. The state having largest area under Bt cotton in India is

(A) Gujarat (B) Punjab

(C) **Maharashtra** (D) Karnataka

180. The First indigenous variety having Bt gene in India is

(A) NHH-44 (B) **Bikaneri Nerma**
(C) MLS-9142 (D) None of the above

181. The Major company supplying Bt cotton in India in India is

(A) Syngenta (B) **Mahyco/Monsanto**
(C) Bayer (D) Rallis India

182. The First Bt cotton varieties approved for cultivation in India is

(A) MECH-12 (B) MECH-184
(C) MECH-162 (D) **All of the above**

183. The Major company supplying Bt cotton in India is

(A) Syngenta (B) **Mahyco/Monsanto**
(C) Bayer (D) Rallis India

184. The term Microbial control was proposed by

(A) Paul De Bach (B) **Steinhaus**
(C) Augustino Bassi (D) Huffaker

185. The Father of Insect Pathology is

(A) Paul De Bach (B) **Steinhaus**
(C) Augustino Bassi (D) Huffaker

186. The Major company supplying Bt cotton in India in India is

(A) Syngenta (B) **Mahyco/Monsanto**
(C) Bayer (D) Rallis India

187. GEAC has given approval for the commercial cultivation of transgenic cotton in India on

(A) **26 March 2002** (B) 26 April 2002
(C) 26 April 2003 (D) 29 May 2003

188. Delfin is a commercial formulation of

(A) ***Bacillus thuringiensis* var *kurstakii***
(B) *Bacillus thuringiensis* var *israeliensis*
(C) Granulosis virus
(D) Nuclear Polyhedroses virus

189. *Bt* cotton, recently commercialized in India is encoded with which of the following *Bt* gene

(A) Cry IAb (B) **Cry IAc**
(C) Cry IIB (D) Cry IIA

190. *Bt* delta endotoxins, after activation bind with receptors of
(A) **Midgut epithelium** (B) Foregut
(C) Trachea (D) Hindgut

191. Caterpillar wilt/Tree top disease or "Wipfelkrankeit" is caused by
(A) **NPV** (B) GV
(C) CPV (D) All of these

192. Grasserie disease of *Bombyx mori* is caused by
(A) **NPV** (B) GV
(C) CPV (D) All of these

193. The genetic material of NPV is
(A) Proteins (B) **DNA**
(C) RNA (D) All of these

194. The Malayan disease in *Oryctes rhinoceros* is caused by
(A) GV (B) **NPV**
(C) **Non occluded virus** (D) None of the above

195. The mode of action of NPV is
(A) **Stomachpoison** (B) Contact poison
(C) Systemic poison (D) None of these

196. The white muscardine fungus is
(A) *Metarhizium anisopliae* (B) ***Beauveria bassiana***
(C) *Verticillium lecanii* (D) *Nomurea rileyi*

197. The Green muscardine fungus is
(A) ***Metarhizium anisopliae*** (B) *Beauveria bassiana*
(C) *Verticillium lecanii* (D) *Nomurea rileyi*

198. The Yellow muscardine fungus is
(A) *Metarhizium anisopliae* (B) *Beauveria bassiana*
(C) ***Apsergillus flavus*** (D) *Sorosporella uvella*

199. The Red muscardine fungus is
(A) *Metarhizium anisopliae* (B) *Beauveria bassiana*
(C) ***Apsergillus flavus*** (D) *Sorosporella uvella*

200. The White-Halo fungus is
(A) *Metarhizium anisopliae* (B) *Beauveria bassiana*
(C) ***Verticillium lecanii*** (D) *Nomurea rileyi*

201. The White-Halo fungus is

(A) *Metarhizium anisopliae* (B) *Beauveria bassiana*
(C) ***Verticillium lecanii*** (D) *Nomurea rileyi*

202. The mode of action of Entomopathogenic fungi is

(A) Stomach poison (B) **Contactpoison**
(C) Systemic poison (D) None of these

203. The mode of action of Protozoan is

(A) **Stomach poison** (B) Contact poison
(C) Systemic poison (D) None of these

204. The Transovarial transmission is observed in

(A) Bacteria (B) Virus
(C) **Protozoa** (D) None of these

205. The mode of action of Protozoan is

(A) **Stomach poison** (B) Contact poison
(C) Systemic poison (D) None of these

206. The Protozoan used for the control of Locust is

(A) ***Nosema locustae*** (B) *Varimorpha necatrix*
(C) *Nosema melolonthae* (D) *Farinocystis tribolii*

207. The Protozoan used for the control of Red flour beetle is

(A) *Nosemalocustae* (B) *Varimorpha necatrix*
(C) *Nosema melolonthae* (D) ***Farinocystis tribolii***

208. The Protozoan used for the control of European corn borer is

(A) *Nosema locustae* (B) ***Varimorpha necatrix***
(C) *Nosema melolonthae* (D) *Farinocystis tribolii*

209. The Protozoan used for the control of chaffer beetles is

(A) *Nosema locustae* (B) *Varimorpha necatrix*
(C) ***Nosema melolonthae*** (D) *Farinocystis tribolii*

210. The Nematodes pathogenic to insects is

(A) *Heterorhabditis indica* (B) *Steinernema carpocapsae*
(C) *Heterorhabditis bacteriophora* (D) **All of the above**

211. The DD-36 is a strain of

(A) *Heterorhabditis indica* (B) ***Steinernema carpocapsae***
(C) *Heterorhabditis bacteriophora* (D) All of the above

212. The entomopathogenic nematode used a biocontrol agent against mosquitoes is

(A) *Heterorhabditis indica* (B) *Steinernema carpocapsae*

(C) *Heterorhabditis bacteriophora* (D) ***Romanomermis culicivorax***

213. The bacterium associated with *Steinernema carpocapsae* is

(A) ***Xenorhabdus* sp.** (B) *Photorhabdus* sp.

(C) Both A and B (D) None of the above

214. The bacterium associated with *Heterorhabditis indica* is

(A) *Xenorhabdus* sp. (B) ***Photorhabdus* sp.**

(C) Both A and B (D) None of the above

215. The infected host turns brick red and luminesces in the dark is the symptom of attack of which nematode

(A) ***Heterorhabditis indica*** (B) *Steinernema carpocapsae*

(C) *Romanomermis culicivorax* (D) All of the above

216. The infective stage of entomopathogenic nematodes is

(A) J1 (B) J2

(C) **J3** (D) J4

217. The rate of application of *Steinernema carpocapsae* for cutworm control is

(A) 2 billion IJs/ha (B) 3 billion IJs/ha

(C) 4 billion IJs/ha (D) **5 billion IJs/ha**

218. The entomopathogenic nematodes kill the hosts in

(A) **24-48 hours** (B) 48-72 hours

(C) 72-96 hours (D) >92 hours

219. The Nemagel is a commercial formulation of

(A) *Heterorhabditis indica* (B) *Steinernema carpocapsae*

(C) *Heterorhabditis bacteriophora* (D) ***Steinernema thermophilum***

220. The NemaStar is a commercial formulation of

(A) *Heterorhabditis indica* (B) ***Steinernema carpocapsae***

(C) *Heterorhabditis bacteriophora* (D) *Steinernema thermophilum*

221. The Ecomask is a commercial formulation of

(A) *Heterorhabditis indica* (B) ***Steinernema carpocapsae***

(C) *Heterorhabditis bacteriophora* (D) *Steinernema thermophilum*

222. The Bio Vector is a commercial formulation of
(A) *Heterorhabditis indica* (B) ***Steinernema carpocapsae***
(C) *Heterorhabditis bacteriophora* (D) *Steinernema thermophilum*

223. The Nemasys is a commercial formulation of
(A) *Heterorhabditis indica* (B) *Steinernema carpocapsae*
(C) *Heterorhabditis bacteriophora* (D) ***Steinernema feltie***

224. The NemaPlus is a commercial formulation of
(A) *Heterorhabditis indica* (B) *Steinernema carpocapsae*
(C) *Heterorhabditis bacteriophora* (D) ***Steinernema feltie***

225. The Etonem is a commercial formulation of
(A) *Heterorhabditis indica* (B) *Steinernema carpocapsae*
(C) *Heterorhabditis bacteriophora* (D) ***Steinernema feltie***

226. The Proactant is a commercial formulation of
(A) ***Steinernema scapterisci*** (B) *Steinernema carpocapsae*
(C) *Heterorhabditis bacteriophora* (D) *Steinernema feltie*

227. The GrubStake is a commercial formulation of
(A) *Steinernemas capterisci* (B) *Steinernema carpocapsae*
(C) ***Heterorhabditis bacteriophora*** (D) *Steinernema feltie*

228. The NemaGreen is a commercial formulation of
(A) *Steinernemas capterisci* (B) *Steinernema carpocapsae*
(C) ***Heterorhabditis bacteriophora*** (D) *Steinernema feltie*

229. The LawnPatrol is a commercial formulation of
(A) *Steinernemas capterisci* (B) *Steinernema carpocapsae*
(C) ***Heterorhabditis bacteriophora*** (D) *Steinernema feltie*

230. The Soil Commandoes is a commercial formulation of
(A) *Steinernemas capterisci* (B) *Steinernema carpocapsae*
(C) ***Heterorhabditis bacteriophora*** (D) *Steinernema feltie*

231. The entomopathogenic nematode enter host through integument
(A) ***Heterorhabditis indica*** (B) *Steinernema carpocapsae*
(C) *Steinernema thermophilum* (D) All of the above

232. The entomopathogenic nematode enter host through integument and natural opening of the host
(A) ***Heterorhabditis indica*** (B) *Steinernema carpocapsae*
(C) Both A and B (D) None of the above

233. The Spicturin is the commercial formulation of

(A) *Bt var. kurstaki* (B) ***Bt var. gallerriae***

(C) *Bt var. tenebrionis* (D) *Bt var. israelensis*

234. The Delfin is the commercial formulation of

(A) ***Bt var. kurstaki*** (B) *Bt var. gallerriae*

(C) *Bt var. tenebrionis* (D) *Bt var. israelensis*

235. The Biobit is the commercial formulation of

(A) ***Bt var. kurstaki*** (B) *Bt var. gallerriae*

(C) *Bt var. tenebrionis* (D) *Bt var. israelensis*

236. The Dipel is the commercial formulation of

(A) ***Bt var. kurstaki*** (B) *Bt var. gallerriae*

(C) *Bt var. tenebrionis* (D) *Bt var. israelensis*

237. The Trident is the commercial formulation of

(A) *Bt var. kurstaki* (B) *Bt var. gallerriae*

(C) ***Bt var. tenebrionis*** (D) *Bt var. israelensis*

238. The Novodor is the commercial formulation of

(A) *Bt var. kurstaki* (B) *Bt var. gallerriae*

(C) ***Bt var. tenebrionis*** (D) *Bt var. israelensis*

239. The M- One is the commercial formulation of

(A) *Bt var. kurstaki* (B) *Bt var. gallerriae*

(C) ***Bt var. tenebrionis*** (D) *Bt var. israelensis*

240. The Sketal is the commercial formulation of

(A) *Bt var. kurstaki* (B) *Bt var. gallerriae*

(C) *Bt var. tenebrionis* (D) ***Bt var. israelensis***

241. The Teknar is the commercial formulation of

(A) *Bt var. kurstaki* (B) *Bt var. gallerriae*

(C) *Bt var. tenebrionis* (D) ***Bt var. israelensis***

242. The Doom is the commercial formulation of

(A) *Bt var. aizawai* (B) ***Bacillus popillae***

(C) *Bacillus sphaericus* (D) *Bt var. sandiego*

243. The Larvicide is the commercial formulation of

(A) *Bt var. aizawai* (B) *Bacillus popillae*

(C) ***Bacillus sphaericus*** (D) *Bt var. sandiego*

244. The Spherimos is the commercial formulation of
(A) *Bt var. aizawai* (B) *Bacillus popillae*
(C) ***Bacillus sphaericus*** (D) *Bt var. sandiego*

245. TheVectobac is the commercial formulation of
(A) *Bt var. aizawai* (B) *Bacillus popillae*
(C) ***Bacillus sphaericus*** (D) *Bt var. sandiego*

246. The Certan is the commercial formulation of
(A) ***Bt var. aizawai*** (B) *Bacillus popillae*
(C) *Bacillus sphaericus* (D) *Bt var. sandiego*

247. The Agree is the commercial formulation of
(A) ***Bt var. aizawai*** (B) *Bacillus popillae*
(C) *Bacillus sphaericus* (D) *Bt var. sandiego*

248. The Xentri is the commercial formulation of
(A) ***Bt var. aizawai*** (B) *Bacillus popillae*
(C) *Bacillus sphaericus* (D) *Bt var. sandiego*

249. The Foil is the commercial formulation of
(A) *Bt var. aizawai* (B) *Bacillus popillae*
(C) *Bacillus sphaericus* (D) ***Bt var. sandiego***

250. The Eclar is the commercial formulation of
(A) **HaNPV** (B) SlNPV
(C) LdNPV (D) *CpGV*

251. The Gypchek is the commercial formulation of
(A) HaNPV (B) SlNPV
(C) **LdNPV** (D) *CpGV*

252. The Madex is the commercial formulation of
(A) HaNPV (B) SlNPV
(C) LdNPV (D) ***CpGV***

253. The Spodopterin is the commercial formulation of
(A) HaNPV (B) **SlNPV**
(C) LdNPV (D) *CpGV*

254. The Mycotrol is the commercial formulation of
(A) ***Beauveria bassiana*** (B) *Metarrhizium anisopliae*
(C) *Verticillium lecanii* (D) *Poeciliomyces fumosoroseus*

255. The Boverin is the commercial formulation of

(A) ***Beauveria bassiana*** (B) *Metarrhizium anisopliae*

(C) *Verticillium lecanii* (D) *Poeciliomyces fumosoroseus*

256. The Bio-Blast is the commercial formulation of

(A) *Beauveria bassiana* (B) ***Metarrhizium anisopliae***

(C) *Verticillium lecanii* (D) *Poeciliomyces fumosoroseus*

257. The Vertilac is the commercial formulation of

(A) *Beauveria bassiana* (B) *Metarrhizium anisopliae*

(C) ***Verticillium lecanii*** (D) *Poeciliomyces fumosoroseus*

258. The Mycotal is the commercial formulation of

(A) *Beauveria bassiana* (B) *Metarrhizium anisopliae*

(C) ***Verticillium lecanii*** (D) *Poeciliomyces fumosoroseus*

259. The PreFeral is the commercial formulation of

(A) *Beauveria bassiana* (B) *Metarrhizium anisopliae*

(C) *Verticillium lecanii* (D) ***Poeciliomyces fumosoroseus***

260. The HaNPV is applied in the field at the rate of

(A) **250-500 LE/ha** (B) 500-75 LE/ha

(C) 750-800 LE/ha (D) None of the above

261. One Larval Equivalent (LE) is

(A) $1x10^6$POB (B) **$1x10^9$ POB**

(C) $1x10^{10}$POB (D) $1x10^{12}$POB

262. The HaNPV is applied in the field at the rate of

(A) 1.5-3.0x 10^{10}POB/ha (B) **1.5-3.0x 10^{12} POB/ha**

(C) 1.5-3.0x 10^9 POB/ha (D) None of the above

263. The Bt is applied in the field at the rate of

(A) **1.00-1.50 kg/ha** (B) 1.50-2.00 kg/ha

(C) 3.00-5.00 kg/ha (D) 20.00 kg/ha

264. The *Paenibacillus popillae* is applied in the field at the rate of

(A) 1.00-1.50 kg/ha (B) **0.50-1.00 kg/ha**

(C) 3.00-5.00 kg/ha (D) 20.00 kg/ha

265. The *Metarrhizium anisopliae* is applied in the field at the rate of

(A) **42.50x 10^{10} spores/m^3** (B) 42.50x 10^{10}spores/ha

(C) $4.50x10^{10}$spores/m^3 (D) None of the above

266. The *Verticillium lecanii* is applied in the field at the rate of

(A) **10 x 10^6 spores/ml** (B) 5 x 10^{12}spores/ml

(C) 10 x 10^{15}spores/ml (D) None of the above

267. The *Beauveria bassiana* can be mass cultured on

(A) Potato Dextrose Aagar medium

(B) Sabouraud's dextrose agar with yeast extract medium

(C) Czapek - Dox agar (CDA) medium

(D) **All of the above**

268. The *Metarrhizium anisopliae* can be mass cultured on

(A) Potato Dextrose Agar medium

(B) Sabouraud's dextrose agar with yeast extract medium

(C) **Emerson's YpSs agar medium**

(D) All of the above

269. The *Paecilomyces fumosoroseus* can be mass cultured on

(A) Oat meal Agar medium

(B) Nutrient Agar medium

(C) Subouraud's maltose agar with yeast extract medium

(D) **All of the above**

270. The *Verticillum lecanii* can be mass cultured on

(A) Oat meal Agar medium

(B) **Potato Dextrsoe Agar medium**

(C) Subouraud's maltose agar with yeast extract medium

(D) All of the above

271. The *Nomuraea rileyi* can be mass cultured on

(A) Oat meal Agar medium

(B) Potato Dextrsoe Agar medium

(C) **Subouraud's maltose agar with yeast extract medium**

(D) All of the above

272. The weed *Parthenium hysterophorus* can be controlled by

(A) ***Zygogramma bicolorata*** (B) *Cactoblastis cactorum*

(C) *Dactylopius ceylonicus* (D) None of the above

273. The weed *Opuntia vulgaris* was successfully controlled by

(A) *Zygogramma bicolorata* (B) *Cactoblastis cactorum*

(C) ***Dactylopius ceylonicus*** (D) None of the above

274. The *Dactylopius ceylonicus* was imported from which country to control *Opuntia vulgaris* in India

(A) Austarlia (B) USA

(C) Thialand (D) **Sri Lanka**

275. The weed *Opuntia inermis* and *O. stricta* was successfully controlled in Australia by

(A) *Zygogramma bicolorata* (B) ***Cactoblastis cactorum***

(C) *Dactylopius ceylonicus* (D) None of the above

276. The weed Water hyacinth was successfully controlled by

(A) *Neochetina bruchi* (B) *Neochetina eichhorniae*

(C) *Orthogalumna terebrantis* (D) **ll the above**

277. The *Neochetina bruchi and N. eichhorniae* was imported from which country to control Water hyacinth in India

(A) Austarlia (B) **USA**

(C) Thialand (D) Sri Lanka

278. The weed *Salvania molesta* was successfully controlled by

(A) ***Cyrtobagous salviniae*** (B) *Neochetina eichhorniae*

(C) *Orthogalumna terebrantis* (D) *Zygogramma bicolorata*

279. The *Cyrtobagous salviniae* wasimported from which country to control *Salvania molesta* in India

(A) **Australia** (B) USA

(C) Thialand (D) Sri Lanka

280. The Coconut mite, *Aceria guerreronis* is an invasive pest introduced from

(A) Africa (B) Central America

(C) **Mexico** (D) Florida

281.The Spiralling whitefly, *Aleurodicus dispersus* is an invasive pest introduced from

(A) Africa (B) **Central America**

(C) Mexico (D) Florida

282. The Papaya mealybug, *Paracoccus marginatus* is an invasive pest introduced from

(A) Africa (B) Central America

(C) Mexico (D) **Florida**

283. The Coffee berry borer, *Hypothenemus hampei* is an invasive pest introduced from

(A) **Africa** (B) Central America

(C) Mexico (D) Florida

284. The Potato tuber moth, *Phthorimaea operculella* is an invasive pest introduced from

(A) Sri Lanka (B) China

(C) **Italy** (D) Florida

285. The Lantana bug, *Orthezia insignis* is an invasive pest introduced from

(A) **Sri Lanka** (B) Central America

(C) Mexico (D) Florida

286. The San Jose Scale, *Quadraspidiotus perniciosus* is an invasive pest introduced from

(A) Sri Lanka (B) **China**

(C) Italy (D) Florida

Unit-8

Insecticide Toxicology

Q 1. Match the following

1. The book Toxicology of Insecticides is authored by

(A) O' Brien R.D. (B) H.C. L. Gupta

(C) Ishaaya I and Degheele (D) **F. Matsumura**

2. The author of book " *Insecticides Action and Metabolism* " is

(A) **O' Brien R.D.** (B) H.C. L. Gupta

(C) Ishaaya I and Degheele (D) F. Matsumura

3. The author of book "*A Critical Review on the Techniques for Testing Insecticides*" is

(A) O' Brien R.D. (B) **J. R. Busvine**

(C) Ishaaya I and Degheele (D) F. Matsumura

4. The book "*Insecticides with Novel Modes of Action*" is authored by

(A) O' Brien R.D. (B) H.C. L. Gupta

(C) **Ishaaya I and Degheele** (D) F. Matsumura

5.The book " *Handbook of Pesticide Toxicology*" is authored by

(A) O' Brien R.D. (B) **W.J. Hays and E. R. Laws**

(C) Ishaaya I and Degheele (D) F. Matsumura

6. Institute of Pesticide Formulation Technology is located at

(A) Faridabad (B) **Gurgaon**

(C) Bangaluru (D) Pune

7. Central Insecticide Laboratory is located at

(A) **Faridabad** (B) Gurgaon

(C) Bangaluru (D) Pune

8. The Headquarter of Central Insecticide Board and Registration Committee is located at
   (A) **Faridabad** (B) Gurgaon
   (C) Bangaluru (D) Pune

9. The All India Coordinated project on Pesticides Residue is located at
   (A) **New Delhi** (B) Gurgaon
   (C) Bangaluru (D) Pune

10. The production of Technical Grade pesticides in India in 2014-15
   (A) 55,540 metric tones (B) **57,353 metric tones**
   (C) 52,980, metric tones (D) 45,390 metric tones

11. Pesticide Industry of India ranks in the world is
   (A) 1st (B) 6th
   (C) 10th (D) **12th**

12. Pesticide Industry of India ranks in Asia is
   (A) **1st** (B) 6th
   (C) 10th (D) 12th

13. The per cent usage of pesticides in India compared to world is
   (A) 1per cent (B) **3 per cent**
   (C) 6 per cent (D) 9 per cent

14. The food grains go waste because of biotic stress is
   (A) **18 per cent** (B) 25 per cent
   (C) 20 per cent (D) 10 per cent

15. The per cent share of insecticides in pesticide use in India is around
   (A) 25 per cent (B) 50 per cent
   (C) **65 per cent** (D) 85 per cent

16. The per cent share of insecticides in pesticide use in the World is around
   (A) **25per cent** (B) 30 per cent
   (C) 35 per cent (D) 40 per cent

17. The insecticide group widely dominate the Indian market is
   (A) Organochlorine (B) Carbamates
   (C) Synthetic pyrethriods (D) **Organophosphate**

18. The per cent share of organophosphate in total pesticide use in India is
   (A) 25 per cent (B) **50 per cent**
   (C) 75 per cent (D) 85 per cent

19. The approximate time to develop one pesticide is

(A) 5 years (B) **10 years**

(C) 15 years (D) 2 years

20. The approximate investment to develop one pesticide is

(A) 50 million dollar (B) 100 million dollar

(C) **140 million dollar** (D) 25 million dollar

21. The per hectare consumption of insecticides in India is

(A) 175g (B) 225g

(C) **275g** (D) 325g

22. The crop having highest pesticide usage is

(A) **Cotton** (B) Rice

(C) Apple (D) Sugarcane

23. The state having highest pesticide usage is

(A) Maharashtra (B) Andhra Pradesh

(C) **Gujarat** (D) Tamil Nadu

24. Approximate number of compounds of active ingredients of pesticides registered world-wide

(A) **800** (B) 900

(C) 1000 (D) 1500

25. The technical grade pesticides registered by Registration Committee (RC) in India under section 9(3) of Insecticides Act, 1968 are

(A) 195 (B) **260**

(C) 330 (D) 385

26. The pesticides formulations registered by Registration Committee (RC) in India under section 9(3) of Insecticides Act, 1968 are

(A) 385 (B) 485

(C) **585** (D) 785

27. The term insecticide is derived from

(A) Italic word (B) **Latin word**

(C) Roman word (D) Greek word

28. The word FIFRA stands for

(A) Federal Insecticide, Fungicide and Rodenticide Act

(B) Fungicide Insecticide, Formulation and Regulation Act

(C) Federal Inspection of Fungicide, and Rodenticide Act

(D) None of the above

29. Branch of medical sciences that deals with the nature, properties, effects and detection of poisons.
    (A) **Toxicology** (B) Insecticide toxicology
    (C) Chemical toxicology (D) None of the above

30. The insecticide labelled as "wonder insecticide" is
    (A) Endoslfan (B) Carbaryl
    (C) **DDT** (D) Malathion

31. The full form of DDT is
    (A) 1,1-Dichloro-2,2-bis(p-chlorophenyl) Methane
    (B) **1, 1 , 1-trichloro-2,2-bis(p-chlorophenyl) ethane**
    (C) 1, 1 , 1-trichloro-2,2-bis (p-chlorophenyl) Methane
    (D) 1,1,1-trichloro-2,2-bis (p-chloromethyyl) ethane

32. The insecticidal properties of DDT was discovered by
    (A) Faraday (B) **Paul Muller**
    (C) Othmar Zeidler (D) Gerhard Schrader

33. The insecticide labelled as "wonder insecticide" is
    (A) Endoslfan (B) Carbaryl
    (C) **DDT** (D) Malathion

34. The first aerial application of an insecticide done at Ohio (USA) against the catalpa Sphinx Moth.
    (A) Dichloro Dimethyl Trichloroethane
    (B) **Dichloro Diphenyl Trichloroethane**
    (C) Dichloro Diphenyltrichloro methane
    (D) Dichloro Diethyl Trichloromethane

35. The first aerial application of an insecticide done in the year at Ohio (USA) against the catalpa Sphinx Moth.
    (A) 1941 (B) 1931
    (C) 1951 (D) **1921**

36. The first organophosphate insecticides was
    (A) Ethion (B) **Parathion**
    (C) Aldrin (D) Malathion

37. Which on of these is propesticide
    (A) Ethion (B) Parathion
    (C) Aldrin (D) **Malathion**

38. The DDT was synthesized by

(A) **Zeidler (1874)** (B) Zeidler (1898)

(C) Gerhard Schrader (1944) (D) Muller (1939)

39. Paul Muller awarded Nobel Prize for discovering insecticidal property of DDT in

(A) 1954 (B) **1948**

(C) 1944 (D) 1939

40. Rachel Carson's book which created awareness about ill effects of pesticides

(A) **Silent Spring (1962)** (B) Pesticides-Boon or Vane

(C) Falling Spring (1962) (D) None of the above

41. The LD_{50} is usually expressed as

(A) µg/larva (B) µg/adult insect

(C) mg/kg body weight (D) **All the above**

42. The lower the LD_{50} dose

(A) **The more toxic the pesticide**

(B) The toxicity is not affected by LD_{50}

(C) The less toxic the pesticide

(D) None of the above

43. This value is used when the exact dose given to the insect cannot be determined

(A) LD_{50} (B) **LC_{50}**

(C) KT_{50} (D) ED_{50}

44. The toxicity of insecticide to mosquito larvae in water can be expressed as

(A) LD_{50} (B) **LC_{50}**

(C) KT_{50} (D) ED_{50}

45. The toxicity parameter used for evaluating synthetic pyrethroids against insects

(A) LD_{50} (B) LC_{50}

(C) **KT_{50}** (D) ED_{50}

46. The toxicity parameter used for evaluating IGRs against insects

(A) LD_{50} (B) LC_{50}

(C) KT_{50} (D) **ED_{50}**

47. Single exposure toxicity over a short period of time

(A) **Acute toxicity** (B) Chronic toxicity

(C) Dermal toxicity (D) Oral toxicity

48. On the basis of **LD_{50}** highly dangerous pesticides are grouped under
   (A) Class IA (B) Class IB
   (C) Class II (D) Class III

49. This method is used when knowledge of the exact amount of insecticide inside the body of the insect is required.
   (A) **Injection method** (B) Residual method
   (C) Oral feeding method (D) Dipping method

50. The Arsenicals are
   (A) **Protoplasmic poison** (B) Physical poison
   (C) Respiratory poison (D) Nerve poison

51. The Biomagnification of pesticides is highest in
   (A) Lower level (B) Middle level
   (C) **Terminal level** (D) All the above

52. The typical drawback of DDT was
   (A) Persistence (B) Insecticide resistance
   (C) Biomagnification (D) **All the above**

53. The first case of DDT resistance was reported in
   (A) House fly (B) Whitefly
   (C) **Singhara beetle** (D) All the above

54. The First full scale use of DDT was in
   (A) World War-I in Italy (B) **World War-II in Italy**
   (C) India Pakistan 1971 War (D) All the above

55. The Ultra low volume concentrates (ULV) solvent volume is
   (A) < 1 Litre (B) **1-5 Litre**
   (C) 5-10 Litre (D) 10-15 per cent

56. The optimum droplet size of insecticidal formulation for flying insects
   (A) **10-15μ** (B) 30-50μ
   (C) 60-120μ (D) 250-500μ

57. The droplet size of aerosol is
   (A) **<50μm** (B) 51-100μm
   (C) 101-200μm (D) >250μm

58. The nozzle type used for herbicide application
   (A) Cone jet nozzle (B) Flat fan
   (C) **Flood jet** (D) Swirl nozzle

59. The nozzle type used for Band application

(A) Cone jet nozzle (B) **Flat fan**

(C) Flood jet (D) Swirl nozzle

60. The nozzle type used for insecticide application

(A) **Hollow Cone nozzle** (B) **Flat fan**

(C) Flood jet (D) Swirl nozzle

61. The insecticides should be sprayed when the insect-pest cross

(A) EI level (B) **ET Level**

(C) Economic Damage (D) Damage Boundary

62. The insecticidal formulation highly toxic to beneficial insects

(A) EC (B) **Dust**

(C) WP (D) SP

63. The efficacy of a pesticide in any application technique is mainly influenced by

(A) Mean level of deposit (B) Distribution of deposit

(C) Wetting agent (D) **All of the above**

64. When pesticide concentration is expressed as not water soluble but soluble in organic solvents

(A) Concentrate (B) **Emulsifiable Concentrate (EC)**

(C) Soluble Concentrates (D) Wettable Powder

65. The concentration of dust formulations range between

(A) **0.1% - 25%** (B) 5-25%

(C) 1-10% (D) 5-10%

66. The Most of insecticides formulations are applied as

(A) **Sprays** (B) Dust

(C) Granular application (D) All the above

67. The characteristics of an appropriate insecticides formulation is/are

(A) Protoplasmic poison (B) Stable and long lasting

(C) Not repellent or irritant to target pests (D) **All the above**

68. The In DDT poisoned insects; there are tremors throughout the body and the appendages characteristically labelled as

(A) Ataxia (B) Hyperactivity

(C) **DDT-Jitter** (D) All the above

69. Carbonic anhydrase, the enzyme responsible for egg shell formation in birds is affected by
   (A) Chlordane (B) **DDT**
   (C) Carbofuran (D) Heptachlor

70. The Insecticidal potency of DDT shows
   (A) Positive temperature dependence
   (B) **Negative temperature dependence**
   (C) temperature independence
   (D) None of the above

71. The Structure – toxicity relationship of DDT was given by
   (A) S. Pradhan (B) **Mullins**
   (C) Beck (D) Matsumura

72. The Insecticidal potency of BHC/Lindane shows
   (A) Positive temperature dependence
   (B) **Negative temperature dependence**
   (C) temperature independence
   (D) None of the above

73. The HCH/BHC was first prepared by
   (A) Van der Lindane (B) **Faraday**
   (C) Gerhard Schrader (D) Othnar Zeidler

74. Benzene hexachloride (BHC) is a misnomer for lindane and other isomers. Its correct name is
   (A) **HCH** (B) GCH
   (C) DDT (D) Lindane

75. Which isomer of Lindane is more toxic
   (A) α-isomer (B) β -isomer
   (C) **γ-isomer** (D) δ-isomer

76. Houseflies poisoned with lindane show
   (A) **Abnormal Fanning movements of their wings** (B) Ataxia
   (C) Prostration (D) Paralysis

77. The Insecticidal potency of Cyclodienes shows

(A) **Positive temperature dependence**
(B) Negative temperature dependence
(C) temperature independence
(D) None of the above

78. The most persistent cyclodiene insecticide is

(A) Aldrin
(B) **Dieldrin**
(C) Ethion
(D) Endosulfan

79. The Pioneer in Organophosphate development was

(A) **Gerhard Scharder**
(B) Philip Clermont
(C) Jean Louis Lassaigne
(D) Willy Lange and Gerde von Krueger

80. The cholinergic nervous system effects of organophosphates was first reported by

(A) Gerhard Scharder
(B) Philip Clermont
(C) Jean Louis Lassaigne
(D) **Willy Lange and Gerde von Krueger**

81. The Sarin, Tabun and Soman are

(A) Organophosphates
(B) Synthetic pyerthroids
(C) **Nerve gases**
(D) Growth Inhibitors

82. The organophosphates was first used against

(A) San Jose Scale
(B) **Colorado potato beetle**
(C) *Helocoverpaarmigera*
(D) Coddling moth

83. The Organophosphates are

(A) acetylcholine inhibitor
(B) GABA inhibitor
(C) **Axonic inhbitor**
(D) All the above

84. The Organophosphates react with the enzyme acetylcholinesterase

(A) **Irreversibly**
(B) Reversibly
(C) Both A and B
(D) None of the above

85. The enzyme system mainly responsible for activation of organophosphate insecticides in insects

(A) Amaylase
(B) **MFO**
(C) Protease
(D) Reductase

86. In Organophospahtes, the process in which rate of recovery of the enzyme become less and less as the time of enzyme inhibitor contact becomes longer in the absence treatment with oximes is
   (A) **Aging** (B) Recovery rate
   (C) Reactivation (D) None of the above
87. The Crabamates are synthetic derivatives of physostigmine (Eserine) which is principal alklaloid of the plant
   (A) *Physostigma venenosum* (B) Calaber beans
   (C) **Both A and B** (D) None of the above
88. The carbamates are synthetic derivatives of
   (A) Physostigmine (B) Esrine
   (C) **Both A and B** (D) None of the above
89. The first successful carbamate insecticide was
   (A) Carbosulfan (B) **Carbaryl**
   (C) Carbofuran (D) Ethion
90. The Carbamates react with the enzyme acetylcholinesterase
   (A) Irreversibly (B) **Reversibly**
   (C) Both A and B (D) None of the above
91. The properties that make synthetic pyrethriods useful for the control of stored product insect-pests are
   (A) High toxicity to insects (B) Favourable toxicity to consumers
   (C) Sufficient persistence (D) **All of the above**
92. The first synthetic pyrethroid was
   (A) Resmethrin (B) **Allethrin**
   (C) Tetramethrin (D) Dimethrin
93. The synthetic pyrethroids are
   (A) **Axonic poison** (B) Presynaptic poison
   (C) Post synaptic poison (D) GABA inhibitors
94. Depending on their mode of action, pyrethroids are categorised as
   (A) **Type 1 and Type 2** (B) Type 1, 2 and 3
   (C) Type 1, 2, 3 and 4 (D) None of the above
95. The type 1 pyrethroids shows
   (A) Positive temperature coefficient (B) **Negative temperature coefficient**
   (C) Both A and B (D) None of the above

96. The Type 2 pyrethroids shows

(A) **Positive temperature coefficient**
(B) Negative temperature coefficient
(C) Both A and B
(D) None of the above

97. Which one of the following is soil active pyrethroid

(A) Resmethrin
(B) Deltamethrin
(C) **Tefluthrin**
(D) Fenvalerate

98. The Tickfugal effect is characteristic of

(A) Organophospahte
(B) Neonicotinoids
(C) **Formamidines**
(D) Avermectins

99. The Formamidines are termed as Insectistatics as they affect

(A) **Locomotion**
(B) Feeding
(C) Reproduction
(D) **All of the above**

100. The Formamidines act as

(A) Synaptic poison
(B) GABA inhibitor
(C) **Octopamine stimulator**
(D) Axonic poison

101. The representative of benzoyl phenyl Urea is

(A) Precocene
(B) **Chitin synthesis inhibitors**
(C) Nitroguanidine
(D) None of the above

102. The First chitin synthesis inhibitor was

(A) **Triflumuron**
(B) Chlorflurazon
(C) Novaluron
(D) Lufenuron

103. The benzoyl phenylurea are

(A) **Stomach poison**
(B) Contact poison
(C) Systemic poison
(D) None of the above

104. The Polyoxins are antibiotics derived from

(A) *Streptomyces cacaoi*
(B) *Streptomyces tendai*
(C) **Both A and B**
(D) None of the above

105. The benzoyl phenylurea exhibits

(A) Translaminar activity
(B) Systemic activity
(C) fumigant activity
(D) **All of the above**

106. Buprofezin, a chitin synthesis inhibitor is marketed under the trade name of

(A) Proclaim
(B) **Applaud**
(C) Mitigate
(D) Fame

107. The Ecdysone (Molting harmone) secreted by
(A) Corpora Cardiaca (B) **Prothoracic glands**
(C) Corpora Allata (D) Neurosecretory cells

108. The JH (neotinin) secreted by
(A) Corpora Cardiaca (B) Prothoracic glands
(C) **Corpora Allata** (D) Neurosecretory cells

109. The Juvenoids/JH mimics were first identified by
(A) **Williams and Slama Karl Slama** (B) V. B. Wigglesworth
(C) Snodgrass (D) Flemmings

110. The 'paper factor' or 'juvabione' was given by
(A) **Carol Williams and** (B) V. B. Wigglesworth
(C) Snodgrass (D) Flemmings

111. The Juvenoids are
(A) Ovicidal (B) larvicidal
(C) **Both A and B** (D) None of the above

112. The JH mimic approved by WHO for use in drinking water to control mosquito larvae
(A) Hydroprene (B) **Methoprene**
(C) Kinoprene (D) Fenoxycarb

113. For insect control purpose, JH activity depends on
(A) The particular insect species (B) Critical phase of insect's life cycle
(C) The mode of administration (D) **All of the above**

114. The Juvenoids affecting larval and pupal development in *Culex pipiens*
(A) **Cyromazine** (B) Fenoxycarb
(C) Tebufenozide (D) Kinoprene

115. The first ecdysone agonist commercialized was
(A) Amethoprin (B) RH 5849
(C) **Tebufenozide** (D) RH 2622

116. The example of Precocene/Anti JH is
(A) Tebufenozide (B) EMD
(C) PB (D) **All of the above**

117. The avermectins are effective against
(A) Mites (B) Insect-pests
(C) Helminths (D) **All of the above**

118. The avermectins are produced by
(A) *Streptomyces cacaoi* (B) *Streptomyces tendai*
(C) ***Streptomyces avermitilis*** (D) None of the above

119. The examples of avermectins are
(A) Abamectin (B) Emamectin
(C) Milbamectin (D) **All of the above**

120. The trade name of abamectin is
(A) Vertimac (B) Avid
(C) Agrimec (D) **All of the above**

121. Emamectin benzoate is sold under the trade name of
(A) Vertimac (B) Avid
(C) Agrimec (D) **Proclaim**

122. The avermectin used against veterinary and medical pests
(A) Abamectin (B) **Ivermectin**
(C) Milbamectin (D) All of the above

123. The avermectins act as
(A) Synaptic poison (B) **GABA inhibitor**
(C) Octopamine stimulator (D) Axonic poison

124. The Spinosadis extracted from actinomycetes
(A) *Streptomyces cacaoi* (B) ***Saccharopolyspora spinosa***
(C) *Streptomyces avermitilis* (D) None of the above

125. The Spinosadis sold under the trade name of
(A) Vertimac (B) Applaud
(C) Proclaim (D) **Tracer**

126. The Milbemycinis extracted from actinomycetes
(A) *Streptomyces cacaoi* (B) *Saccharopolyspora spinosa*
(C) *Streptomyces avermitilis* (D) ***Streptomyces hygroscopicus* subsp. *aureolacrimosus***

127. The Polynactins is produced by
(A) ***Streptomyces aureus*** (B) *Saccharopolyspora spinosa*
(C) *Streptomyces avermitilis* (D) None of the above

128. The Polynactins act as
(A) Synaptic poison (B) GABA inhibitor
(C) Octopamine stimulator (D) **Uncoupler of oxidative Phosphorylation**

129. The synthetic analogues of nicotine are known as

(A) Nicotinoids (B) **Neonicotinoids**
(C) Synthetic nicotinoids (D) None of the above

130. The neonictinoids recommended for seed treatment

(A) Imidacloprid 70 WS (B) Thiamethoxam 70 WS
(C) Both A and B (D) None of the above

131. The Neonicotinoids act as

(A) **Agonists at the nAChR** (B) GABA inhibitor
(C) Uncoupler (D) Axonic poison

132. The neonictinoids recommended for seed treatment

(A) Imidacloprid70 WS (B) Thiamethoxam 70 WS
(C) **Both A and B** (D) None of the above

133. Acetamiprid is a

(A) **Neonictinoid** (B) Organophosphate
(C) Pyrethroid (D) Ecdysteroid

134. The example of pyridincarboxamide is

(A) **Fluconamid 50 WG** (B) Thiamethoxam 70 WS
(C) Dinotefuran 20 SG (D) None of the above

135. The characteristics of pyridincarboxamide are

(A) Inhibit feeding behaviour (B) Translaminar activity
(C) Good persistence (D) **All of the above**

136. The pyridincarboxamide are effective against

(A) Aphids (B) Whiteflies
(C) Mealybugs (D) **All of the above**

137. The Phenyl Pyrazoles act as

(A) Agonists at the nAChR (B) **GABA inhibitor**
(C) Uncoupler (D) Axonic poison

138. The example of Phenyl Pyrazoles is

(A) Fipronil (B) Pyriprole
(C) Ethiprole (D) **All of the above**

139. The example of Pyridine azomethines is

(A) **Pymetrozine** (B) Pyriprole
(C) Ethiprole (D) All of the above

140. The characteristic of Pyridine azomethines is
   (A) Block stylet penetration causing immediate cessation of feeding of insects
   (B) Effective against sucking pests
   (C) Can be applied as foliar and soil application
   (D) **All of the above**

141. The example of Oxadiazone is
   (A) Fipronil (B) Pyriprole
   (C) **Indoxacarb** (D) Fipronil

142. The indoxacarb is sold under the trade name of
   (A) Chess (B) **Avaunt**
   (C) Regent (D) Toekn

143. The example of halogenated pyrroles is
   (A) Fipronil (B) Pyriprole
   (C) **Chlorfenpyr** (D) All of the above

144. The trade name of Chlorfenpyr is
   (A) **Interprid** (B) Pyriprole
   (C) Ethiprole (D) All of the above

145. The example of Thiazolidine group is
   (A) Fipronil (B) Pyriprole
   (C) **Hexythiazox** (D) All of the above

146. Hexythiazox is sold under the trade name of
   (A) Fipronil (B) **Maiden**
   (C) Ethiprole (D) All of the above

147. Hexythiazox is an effective
   (A) **Acaricide** (B) Insecticide
   (C) Fungicide (D) Rodenticide

148. The example of Thiourea derivative is
   (A) Fipronil (B) Maiden
   (C) **Diafenthiuron** (D) All of the above

149. Diafenthiuron is sold under the trade name of
   (A) **Polo** (B) Maiden
   (C) Ethiprole (D) Magister

150. The mode of action of Diafenthiuron is

(A) Axonic Poison (B) Synaptic poison

(C) GABA inhibitor (D) **Inhibit oxidative phosphorylation**

151. The Propargite, an acaricide belong to

(A) Thiourea derivative group (B) Avermectin

(C) **Sulfite Ester group** (D) Diamide group

152. The mode of action of propargite is

(A) Axonic Poison (B) Synaptic poison

(C) GABA inhibitor (D) **Inhibit oxidative phosphorylation**

153. The trade name of Propargite is

(A) **Omite** (B) Maiden

(C) Magister (D) None of the above

154. The insecticide Flubendamide and Chlorantraniliprole belong to

(A) Thiourea derivative group (B) Avermectin

(C) Sulfite Ester group (D) **Diamide group**

155. The Flubendamide and Chlorantraniliprole act as

(A) **Inhibit Ryanodine receptor** (B) Synaptic poison

(C) GABA inhibitor (D) Inhibit oxidative phosphorylation

156. The trade name of Flubendamide is

(A) Fame (B) Takumi

(C) Magister (D) **Both A and B**

157. The trade name of Chlorantraniliprole is

(A) Ferterra (B) Coragen

(C) **Both A and B** (D) Magister

158. The acaricide, fenzaquin belong to

(A) **Quniazoline group** (B) Diamide group

(C) Avermectin (D) Thiazolidine group

159. The mode of action of Chlorantraniliprole is

(A) **Inhibit Electorn transport chain** (B) Synaptic poison

(C) GABA inhibitor (D) Inhibit oxidative phosphorylation

160. Fenzaquin is sold under the trade name of

(A) Ferterra (B) Coragen

(C) Fame (D) **Magister**

161. Spiromesifen, an insecticide cum acaricide belong to

(A) Thiazolidine group (B) **Tetronic acid derivative**

(C) Avermectin (D) Diamide group

162. The mode of action of Spiromesifen is

(A) **Inhibit Lipid synthesis** (B) Synaptic poison

(C) GABA inhibitor (D) Inhibit oxidative phosphorylation

163. Fenzaquin is sold under the trade name of

(A) Ferterra (B) Coragen

(C) Fame (D) **Magister**

164. Bifenazate, an carbazate acaricide is sold under the trade name of

(A) Ferterra (B) Coragen

(C) **Floramite** (D) Magister

165. Fenpyroximate is an

(A) Insecticide (B) Nematicide

(C) Fumigant (D) **Acaricide**

166. Fenpyroximate, an Pyridazinone acaricide is sold under the trade name of

(A) **Mitigate** (B) Coragen

(C) Fame (D) Magister

167. The mode of action of Fenpyroximate is

(A) Inhibit Lipid synthesis (B) Synaptic poison

(C) GABA inhibitor (D) **Inhibit Electron transport**

168. Neiristoxin is poison from a marine annelid

(A) *Lumbrineiris brevicirra* (B) *Lumbrineiris heteropoda*

(C) **Both A and B** (D) None of the above

169. Synthetic derivatives of neiristoxin are

(A) Cartap hydrochloride (B) Bensultap

(C) Thiocyclam (D) **All of the above**

170. Fumigant known as tear gas is

(A) **Chloropicrin** (B) Carbon Tetrachloride

(C) p- dichlorobenze (D) Dibromochloropropane

171. Increase in the bioactivity of two compounds to produce an effect greater than the one expected from a single algebraic summation of the effects of two compounds individually administered is known as

(A) Activation (B) Synergism

(C) Potentiation (D) **Both A and B**

172. The term Potentiation is used where

(A) **Both the compounds are toxic** (B) One compounds is toxic

(C) Both the compounds are Non toxic (D) None of the above

173. The Synergist which markedly increases the activity of insecticides are

(A) Piperonyl buroxide (B) MGK264

(C) Sesamolin (D) **All the above**

174. The term synergist enhance the insecticide activity by

(A) Inhibition of mixed function oxidase enzymes/ hydrolyzing enzymesc

(B) Probably serving as alternative substrate

(C) Two compounds interact with different sites of the target

(D) **All the above**

175. The Rotenone are isolated chemical from and named it as Rotenone

(A) **Roots of *Derris*** sp. (B) Bark of *Derris* sp.

(C) Leaves of *Derris* sp. (D) All of the above

176. The Flavonoids Rotenone are extracted from roots of

(A) *Lonchocarpus* sp. (B) *Derris malaccensis*

(C) *Derris elliptica* (D) **All of the above**

177. The mode of action of Rotenone is

(A) Inhibit Lipid synthesis (B) Synaptic poison

(C) GABA inhibitor (D) **Inhibit Electron transport**

178. Sabadilla, an alkaloid is obtained from the

(A) Leaves (B) Bark

(C) **Seeds** (D) Roots

179. Sabadilla, an alkaloid is obtained from the ripe seeds of

(A) ***Schoenocaulon officinale*** (B) *Nicotiana tabacum*

(C) *Ageratum* sp. (D) *Azadirachta indica*

180. Insecticidal activity of sabadilla comes from the alkaloid fraction of

(A) Veratridine (B) Cevadine

(C) **Both A and B** (D) None of the above

181. The mode of action of sabadilla is similar that of

(A) Carbamates (B) Organophosphate

(C) Neonictinoids (D) **Synthetic Pyerthroids**

182. Sabadilla is highly toxic to which of the non target organism

(A) **Honeybee** (B) *Chrysoperla carnea*

(C) Lady bird beetle (D) All the above

183. Nicotine is an alkaloid obtained from

(A) *Nicotiana Rustica* (B) *N. tabacum*

(C) **Both A and B** (D) None of the above

184. The pure Nicotine alkaloid was first isolated by

(A) Berliner (B) **Posselt and Reimann**

(C) Zeidler (D) None of the above

185. Nicotine First synthesised in 1904 under trade name

(A) Green Leaf 40 (B) Nicotine 40

(C) **Black Leaf 40** (D) None of the above

186. Nicotine is

(A) Contact Poison (B) Stomach Poison

(C) Both A and B (D) None of the above

187. The mode of action of nicotine is

(A) Pre synaptic aectylcholine inhibitor (B) Axonic poison

(C) **Post synaptic aectylcholine inhibitor** (D) GABA inhibitor

188. The synthetic derivative of nicotine are known as

(A) Neo-Nicotinoids (B) Nitroguanidine

(C) **Both A and B** (D) None of the above

189. The Azadirachtin is a tetraterpenoid compound found in

(A) Leaves (B) Fruits

(C) Seeds (D) **All of the above**

190.T he highest cocentration of Azadirachtin is found in

(A) Leaves (B) Fruits

(C) **Seeds** (D) All of the above

191. Azadirachtin, an triterpenoid is obtained from the seeds of

(A) *Schoenocaulon officinale* (B) *Nicotiana tabacum*

(C) *Melia azedrach* (D) ***Azadirachta indica***

192. The Neem seed contains

(A) Azadirachtin (B) Salannin

(C) Meliantriol (D) **All of the above**

193. The main constituent of neem tetrano triterpeniod is

(A) **Azadirachtin** (B) Salannin

(C) Meliantriol (D) All of the above

194. Of the total tetrano triterpeniod in neem seed, Azadirachtin constitutes

(A) 70 per cent (B) 80 per cent

(C) **90 per cent** (D) 5-10 per cent

195. The mode of action of neem is

(A) Repellent (B) Growth regulating effect

(C) Antifeedant (D) **All of the above**

196. The commercial formulation of neem is

(A) Neemcure (B) Neemark

(C) Neemazal (D) **All of the above**

197. Pyrethrins are extracted from the daisy-like flower of

(A) ***Chrysanthemum cinerariaefolium*** (B) *Sabadilla* sp

(C) *Ryania speciosa* (D) *Schoenocaulon officinale*

198. The pyrethrins are extracted from the which part of *Chrysanthemum*

(A) Leaves (B) **Flower**

(C) Seed (D) All of the above

199. Pyrethrum is predominantly found in which of the following parts of *Chrysanthemum*

(A) Leaves (B) Achenes of **Flower**

(C) Seed (D) Petals of flower

200. A built-in tank is absent in

(A) Hand spayer (B) Power sprayer

(C) **Rocker sprayer** (D) Knapsack sprayer

201. When the *C. cinerariaefolium* flower is ground into a powder, the product is

(A) Pyrethrin (B) **Pyrethrum**

(C) Pyrethroid (D) All of the above

202. The active compounds of *C. cinerariaefolium* flower is

(A) **Pyrethrin** (B) Pyrethrum

(C) Pyrethroid (D) All of the above

203. Synthetic insecticides that mimic the action of the pyrethrins are known as
(A) Pyrethrin (B) Pyrethrum
(C) **Pyrethroid** (D) All of the above

204. Pyrethrins are six esters formed by the combination of the
(A) **Acids and Alcohol** (B) Acid and alkalis
(C) Acid and Phenols (D) Acids and Ester

205. Pyrethrins – I are esters of
(A) **Chrysenthemic acid+Pyrethrolone**
(B) Chrysenthemic acids + Jasmolone
(C) Chrysenthemic acids+Cinerolone
(D) Pyrethric acid+Jasmolone

206. Cinerins – I are esters of
(A) Chrysenthemic acid+Pyrethrolon
(B) Chrysenthemic acids + Jasmolone
(C) **Chrysenthemic acids+Cinerolone**
(D) pyrethric acid+Jasmolone

207. Jasmolin–I are esters of
(A) Chrysenthemic acid+Pyrethrolone
(B) **Chrysenthemic acids + Jasmolone**
(C) Chrysenthemic acids+Cinerolone
(D) pyrethric acid+Jasmolone

208. Pyrethrins-II are esters of
(A) **Pyrethric acid+Pyrethrolone** (B) Pyrethric acid + Jasmolone
(C) Pyrethric acid +Cinerolone (D) Chrysenthemic acid+Pyrethrolone

209. Cinerin-II are esters of
(A) Pyrethric acid+Pyrethrolone (B) Pyrethric acid + Jasmolone
(C) **Pyrethric acid +Cinerolone** (D) Chrysenthemic acid+Pyrethrolone

210. Jasmolone-II are esters of
(A) Pyrethric acid+Pyrethrolone (B) **Pyrethric acid + Jasmolone**
(C) Pyrethric acid +Cinerolone (D) Chrysenthemic acid+Pyrethrolone

211. The cockroaches affected by pyrethrum spray typically shows

(A) Knockdown effect (B) **Flushing effect**

(C) Both A and B (D) None of the above

212. The Pyrethrums shows

(A) Positive temperature coefficient (B) **Negative temperature coefficient**

(C) Both A and B (D) One of the above

213. The Pyrethrums are most effective against

(A) Foliage feeding insects (B) Sucking insect

(C) **Flying insects** (D) All the above

214. The synergists usually mixed with pyrethrums is/are

(A) Sesamin (B) Sesamolin

(C) Piperonyl butoxide (D) **All the above**

215. The insecticide group used in aircrafts for control of insects as quarantine measure

(A) Carbamates (B) **Pyrethrums**

(C) Organophosphates (D) Pyrethroids

216. The formulation of pyrethrum used in international aircrafts for control of insects

(A) Spray (B) Dust

(C) **Aerosol** (D) Fumigant

217. The Ryania compound is obtained from thewhich part of the plant

(A) Stem (B) Root

(C) **Both A and B** (D) Seed

218. The Ryania compound is obtained from the roots and stems of which plant

(A) ***Ryania speciosa*** (B) *Derris elliptica*

(C) Euclayptus (D) *Melia azedrach*

219. The chemical constituent found in *Ryania speciosa*

(A) Ryanodine (B) Ryanodol

(C) 9,21-dehydroryanodine (D) **All of the above**

220. The most active chemical constituent of *Ryania speciosa*

(A) Ryanodine (B) 9,21-dehydroryanodine

(C) **Both A and B** (D) All of the above

221. The botanical insecticide having longer residual activity than most other botanicals
(A) **Ryania** (B) Neem
(C) Derris (D) Abadilla

222. The In-compatibility types in insecticide
(A) Chemical incompatibility (B) Biological incompatibility
(C) Physical incompatibility (D) **All of the above**

223. The In-compatibility type in which the two pesticides react with the another producing a different compound, reducing the pesticidal activity of the pesticides
(A) **Chemical incompatibility** (B) Biological incompatibility
(C) Physical incompatibility (D) All of the above

224. The In-compatibility types in which the mixed product exhibit phytotoxic action, which independently is not phytotoxic
(A) Chemical incompatibility (B) **Biological incompatibility**
(C) Physical incompatibility (D) All of the above

225. The agglomeration, phase separation, explosive reaction, etc. are which type of In-compatibility in insecticide
(A) Chemical incompatibility (B) Biological incompatibility
(C) **Physical incompatibility** (D) All of the above

226. The disease resulting from ingestion of sea food contaminated by methyl mercury originating from industrial waste known as
(A) Blue baby (B) **Minamata disease**
(C) Both A and B (D) None of the above

227. The Bhopal Gas Tragedy occur in the year
(A) **1984** (B) 1990
(C) 1994 (D) 1996

228. The Bhopal Gas Tragedy is linked with
(A) Endosulfan (B) **Carbaryl**
(C) Parathion (D) Carbofuran

229. The Bhopal Gas Tragedy occur due to the leakage of which gas
(A) Sodium isocyanate (B) Calcium isocyanate
(C) **Methyl isocyanate** (D) Potassium isocyanate

230. The insecticide banned in Kerala for spraying in cashew Plantation
(A) **Endosulfan** (B) Carbaryl
(C) Parathion (D) Carbofuran

231. As per WHO estimates, annually deaths and poisoning cases worldwide are
(A) 10,000 and 1 Lakh (B) **20,000 and 5 Lakh**
(C) 30,000 and 6 Lakh (D) 50,000 and 5 Lakh

232. Rachael carson's *Silent Spring* (1962) pointed which insecticide group responsible for thinning of egg shell or reproductive failure in birds
(A) **Organochlorines** (B) Organophosphates
(C) Carbamates (D) All the above

233. The insecticide formulations more toxic to honeybee
(A) **Dust** (B) Spray
(C) Wettable powder (D) Granules

234. The organochlorines are responsible for
(A) Bio-concentration (B) Bioaccumulation
(C) Bio-magnification (D) **All the above**

235. Accumulation of insecticides at different trophic levels of food chain
(A) Bio-concentration (B) Bioaccumulation
(C) Bio-magnification (D) **All the above**

236. The per cent of the pesticide applied to crop reaches the target.
(A) **1 per cent** (B) 10 per cent
(C) 15 per cent (D) 50 per cent

237. Development of an ability in a given strain insects to tolerate doses of an insecticide which would prove lethal to the other individuals of the same species
(A) Tolerance (B) **Resistance**
(C) Both A and B (D) None of the above

238. Based on pesticide resistance data, highest resistance is reported in
(A) **Insects** (B) Mites
(C) Nematodes (D) Diamond Back Moth

239. The per cent of resistance development in Insects among animal kingdom is
(A) 55 per cent (B) 65 per cent
(C) 75 per cent (D) **85 per cent**

240. The number of insect-pests reported develop insecticide resistance is
(A) >500 (B) >550
(C) **>600** (D) >650

241. The highest number of resistant species are reported from which insect order

(A) **Diptera** (B) Homoptera
(C) Lepidoptera (D) Coleoptera

242. The per cent of resistance development in order Diptera is

(A) 20 per cent (B) 30 per cent
(C) **39 per cent** (D) 45 per cent

243. The first insect pest developed resistance worldwide

(A) **San Jose Scale** (B) Wooly Apple Aphid
(C) *Helicoverpa armigera* (D) Diamond Back Moth

244. The San Jose Scale was reported to develop insecticide resistance against

(A) Endosulfan (B) **Lime sulfur**
(C) DDT (D) Malathion

245. San Jose Scale developed resistance against lime sulphur in Washington in the year

(A) **1908** (B) 1946
(C) 1951 (D) 1971

246. The first insect pest reported develop insecticide resistance in India

(A) San Jose Scale (B) **Singhara beetle**
(C) *Helicoverpa armigera* (D) Diamond Back Moth

247. The Singhara beetle was reported to develop insecticide resistance against

(A) Endosulfan (B) Lime sulfur
(C) **DDT** (D) Malathion

248. Singhara beetle developed resistance against DDT in the year

(A) 1908 (B) **1946**
(C) 1951 (D) 1971

249. In resistance management, the rate of insecticide which kills resistant heterozygotes is referred to as

(A) **Highdose** (B) Low dose
(C) Sublethal dose (D) Refugia

250. Daily intake of a chemical, which during an entirelifetime will not cause any appreciable risk is

(A) Maximum Residue Limit (B) **Acceptable Daily Intake**
(C) Minimum Residue Limit (D) Acceptable chemical Intake

251. Particles suspended in aerosol are of the size ranging from

(A) **0.1-50μ** (B) 50-100μ

(C) 100-200μ (D) 0.65-25μ

252. An increase in population of target insect after insecticidal spray is referred to as

(A) **Resurgence** (B) Hormoligosis

(C) Replacement (D) Secondary pest outbreak

253. Situation where a population, after having been suppressed, rebounds to number greater than before suppression occurred

(A) **Resurgence** (B) Hormoligosis

(C) Replacement (D) Secondary pest outbreak

254. Situation when a major pest is suppressed and continues to be suppressed by a tactic but is replaced but is replaced by another pest, previously with minor status

(A) Replacement (B) Secondary pest outbreak

(C) **Both A and B** (D) Hormoligosis

255. Direct favourable influence of pesticides on physiology and behaviour of insect pest is known as

(A) Resurgence (B) **Hormoligosis**

(C) Replacement (D) Secondary pest outbreak

256. The Increased resistance leads to vicious cycle of increased dosage and frequency of pesticide application

(A) **Pesticide treadmill** (B) Excessive pesticide use

(C) Pesticide Overdose (D) All of the above

257. The financial cost of resistance is estimated on worldwide basis is

(A) **$ 1 billion** (B) $ 5 billion

(C) $ 10 billion (D) $ 15 billion

258. The insecticide group having maximum cases of insecticide resistance

(A) DDT (B) **Cyclodiene**

(C) Organophosphate (D) Synthetic pyrethroids

259. Resistance against selective agent and other compounds is developed by the same mechanism

(A) **Cross resistance** (B) Multiple resistance

(C) Both A and B (D) Non of the above

260. Co-existence of different defence mechanisms in the same strain

(A) Cross resistance (B) **Multiple resistance**

(C) Both A and B (D) None of the above

261. The rate at which resistance develops in population depends on

(A) The frequency and nature of resistant genes present in the population

(B) The intensity of selection pressure

(C) The rate at which population breeds

(D) **All of the above**

262. Factors involved in insect resistance to insecticides include

(A) Target insensitivity (B) Increased rate of excretion

(C) Increased detoxification (D) **All of the above**

263. Coddling moth due to its acquired habit rejects the first bite prior to penetration in the fruit represents

(A) **Behavioural resistance** (B) Physiological resistance

(C) Biochemical resistance (D) All of the above

264. The gene responsible for resistance to DDT in housefly

(A) mpf gene (B) hfr gene

(C) **kdr gene** (D) All of the above

265. The recessive gene kdr (knockdown resistance) in house fly is located on which chromosome

(A) Chromosome 1 (B) **Chromosome 3**

(C) Chromosome 5 (D) Chromosome 7

266. The insecticide resistance can be managed by

(A) Rotation or alteration with different insecticides

(B) Use of multiple components of IPM in combination with pesticides

(C) Preserve refugia and avoid slow release formulations

(D) **All the above**

267. The antidote recommended in organochlorine poisoning is

(A) Phenobarbital (B) Diazepam

(C) Paraldehyde (D) **All the above**

268. The antidote recommended in organophosphate poisoning is

(A) 2-PAM (B) Atropine

(C) Toxogonin (D) **All the above**

269. The antidote recommended in carbamate poisoning is

(A) 2-PAM (B) **Atropine**
(C) Toxogonin (D) All the above

270. The antidote recommended in Synthetic pyrethroids poisoning is

(A) 2-PAM (B) Atropine
(C) **Barbiturates** (D) All the above

271. The antidote recommended in Stomach poisoning in general is

(A) Common salt (B) Magnesium oxide
(C) Activated charcoal (D) **All the above**

272. The antidote recommended in Zinc phosphide poisoning is

(A) Calcium Gluconate (B) **Potassium permanganate**
(C) Methyl Nitrite ampule (D) Epinephrine

273. The antidote recommended in Methyl Bromide poisoning is

(A) Calcium Gluconate (B) Potassium permanganate
(C) Methyl Nitrite ampule (D) **Epinephrine**

274. The antidote recommended in Cyanide poisoning is

(A) Calcium Gluconate (B) Potassium permanganate
(C) **Methyl Nitrite ampule** (D) Epinephrine

275. The phenomenon of conversion of some insecticides in to more toxic substances when enter in the insect body is

(A) **Activation** (B) Potentiation
(C) Toxification (D) Synergism

276. The enzyme system responsible for insecticide metabolism

(A) **Cytochrome P_{450}** (B) Cytochrome P_{100}
(C) Cytochrome P_{250} (D) All of the above

277. The microsomal oxidation concerned with insecticide metabolism

(A) Mixed function oxidase system
(B) Microsomal oxidase system
(C) Monooxygenase function oxidase system
(D) **All of the above**

278. The insecticide Act was implemented in the year

(A) **1968** (B) 1971
(C) 1914 (D) 1984

279. The insecticide Rules was implemented in the year

(A) 1968 (B) **1971**

(C) 1914 (D) 1984

280. The Environment (Protection) Act was implemented in the year

(A) 1968 (B) **1986**

(C) 1964 (D) 1994

281. The Hazardous Waste (Management & Handling) Rules was implemented in

(A) 1968 (B) 1986

(C) 1964 (D) **1989**

282. The Water (Prevention & Control of Pollution) Act was implemented in

(A) 1968 (B) 1986

(C) **1974** (D) 1989

283. The chairman of Central Insecticide Board is

(A) **DG, Health Services** (B) DG, ICAR

(C) DDG, crop Science, ICAR (D) DDG, Health Services

284. The Chairman of Insecticide Registration Committee

(A) DG, Health Services (B) DG, ICAR

(C) **DDG, Crop Science, ICAR** (D) DDG, Health Services

285. The Water (Prevention & Control of Pollution) Act was implemented in

(A) 1968 (B) 1986

(C) **1974** (D) 1989

286. The MRL is fixed by

(A) Ministry of Agriculture

(B) Ministry of Chemicals and fertilizers

(C) **Ministry of Health and Family Welfare**

(D) All of the above

287. The most widely used triazene which inhibit feeding in cockroaches

(A) **AC-24055** (B) Neem

(C) Triphenyl tin acetate (D) All the above

288. Phlorizin (**Apple factor**) extracted from apple is an effective antifeedant

(A) Apple aphids (B) **Non Apple feeding aphids**

(C) Both A and B (D) All the above

289. Commercial mosquito repellent 'Odomos' contain oil extracted from lemongrass, *Andrpogon pardus* as repellent

(A) **Citronella oil** (B) Neemoil
(C) Camphor oil (D) Cedarwood oil

290. The repellent used against tsetse fly, *Glossina morsitans* is

(A) **Citronella oil** (B) Pyrethrum
(C) Camphor oil (D) Cedarwood oil

291. Cuelure is attractant to

(A) *Dacus dorsalis* (B) *Ceratitis capitata*
(C) ***Dacus cucrbitae*** (D) All the above

292. Methyl Eugenol is attractant to

(A) ***Dacus dorsalis*** (B) *Ceratitis capitata*
(C) *Dacus cucrbitae* (D) All the above

293. Trimedlure is attractant to

(A) *Dacus dorsalis* (B) ***Ceratitis capitata***
(C) *Dacus cucrbitae* (D) All the above

294. The lure used to attract and kill cockroaches is

(A) **Sweet syrup + white or yellow phosphorus**
(B) Thallous sulphste + sugar + honey + glycerine +water
(C) Thallous sulphate + peanut butter
(D) All the above

295. The lure used to attract and kill meat loving ants is

(A) Sweet syrup + white or yellow phosphorus
(B) Thallous sulphste + sugar + honey + glycerine +water
(C) **Thallous sulphate + peanut butter**
(D) All the above

296. The lure used to attract and kill sweet-loving ants is

(A) Sweet syrup + white or yellow phosphorus
(B) **Thallous sulphste + sugar + honey + glycerine +water**
(C) Thallous sulphate + peanut butter
(D) All the above

297. Repellents used against Mosquito, blood suckers

(A) Benzyl benzoate (B) Trichlorobenzene
(C) **Dimethyl pthalate** (D) Pentachlorophenol

298. Repellents used against Mites (chiggers)
(A) **Benzylbenzoate** (B) Trichlorobenzene
(C) Dimethylpthalate (D) Pentachlorophenol

299. Repellents used against Crawling insects
(A) Benzyl benzoate (B) **Trichlorobenzene**
(C) Dimethyl pthalate (D) Pentachlorophenol

300. Repellents used against wood feeders
(A) Benzyl benzoate (B) Trichlorobenzene
(C) Dimethyl pthalate (D) **Pentachlorophenol**

301. The insecticide completely banned for use in vegetables
(A) Endosulfan (B) Aluminum Phosphide
(C) Carbaryl (D) **Monocrotophos**

302. The insecticide restricted for use is
(A) Endosulfan (B) **Aluminum Phosphide**
(C) Carbaryl (D) Monocrotophos

303. The insecticide recently banned in India
(A) **Endosulfan** (B) Aluminum Phosphide
(C) Carbaryl (D) Monocrotophos

304. The use of DDT in India is permitted (with restriction) under expert Government supervision in
(A) Agriculture (B) Domestic public health
(C) **Both A and B** (D) None of the above

305. Which of the following crops consumes higher amount of insecticides used for plant protection in India
(A) Sugarcane (B) Vegetables
(C) **Cotton** (D) Rice

306. Scheduled spray of insecticides without considering any economic decision levels were done during
(A) Subsistence phase (B) Crisis phase
(C) **Exploitation phase** (D) Disaster phase

307. Universal antidote is the mixture of activated charcoal : tannic acid : MgO in the ratio of
(A) **2:1:1** (B) 1:2:1
(C) 1:1:2 (D) 1:1:2.5

308. Each Tablet of Celphos consists of

(A) **AlP (56%) + carbamate (44%)** (B) AlP (50%) + carbamate (50%)

(C) AlP (60%) + carbamate (40%) (D) AlP (44%) + carbamate (56%)

Unit-9

Insect Pests of Field Crops

Q 1. Match the following

1. The author of "Insect and mites of crops in India" is

(A) Atwal AS & Dhaliwal GS (B) **MRGK Nair**

(C) Butani DK & Jotwani MG. (D) AR Horowitz and I Ishaya

2. The author of "*Insects and Vegetables*" is

(A) Atwal AS & Dhaliwal GS (B) MRGK Nair

(C) **Butani DK & Jotwani MG.** (D) AR Horowitz and I Ishaya

3. The author of "*Agricultural Pests of South Asia and their Management* " is

(A) **Atwal AS & Dhaliwal GS** (B) MRGK Nair

(C) Butani DK & Jotwani MG. (D) AR Horowitz and I Ishaya

4. Which of the following pathogens is effective against citrus rust mite

(A) *Verticillium lecanii* (B) *Hirsutella thompsonii*

(C) *Nomurea rileyi* (D) ***Nosema cuneatum***

5. 'Silver shoot' in rice produced by gallmidge is a modification of

(A) **Leaf sheath** (B) Leaf tip

(C) Stem (D) Leaf base

6. Eggs of mango mealy bug are laid in

(A) Leaves (B) **Soil**

(C) Crevices (D) Tree trunk

7. Bhindi yellow vein clearing disease is transmitted by

(A) *Oligonychus indicus* (B) ***Bemisia tabaci***

(C) *Aphis gossypii* (D) *Amrasca devestans*

8. *Epiricania melanoleuca,* an ectoparasitoid is effective against

(A) **Sugarcane pyrilla** (B) Sugarcane mealy bug

(C) Sugarcane scale (D) Sugarcane whitefly

9. Predominant predator in the cotton ecosystem
   (A) ***Cheilomenes sexmaculatus*** (B) *Oxycetonia versicolor*
   (C) *Altica cyani* (D) *Zygogramma bicolorata*

10. Gossyplure is an effective sex pheromone used in IPM against
   (A) ***Pectinophora gossypiella*** (B) *Helicoverpa armigera*
   (C) *Spodoptera litura* (D) *Erias vittella*

11. *Euglandina rosae* is a predator employed against
   (A) Castor slug caterpillar (B) Citrus mealybug
   (C) **Giant AfricanSnail** (D) *Parthenium sp.*

12. Mite pest which has become a serious threat to coconut in the recent years is
   (A) *Aceria thompsonii* (B) ***Aceria guerrreronis***
   (C) *Aceria cajani* (D) *Aceria indicus*

13. Following which insect is a pest in adult stage
   (A) ***Otheris fullonica*** (B) *Pieris brassicae*
   (C) *Spodoptera litura* (D) *Helicoverpa armigera*

14. Rosetting of flowers and interlocular burrowing of bolls in cotton is due to
   (A) ***Pectinophora gossypiella*** (B) *Erias vittella*
   (C) *Spodoptera litura* (D) *Helicoverpa armigera*

15. Wooly apple aphid resistant rootstock of apple is
   (A) *M-2* (B) *M-16*
   (C) *M-27* (D) **Northern Spy**

16. The cultivar of apple is good pollinizers for all the Delicious group
   (A) ***Golden Deliciou*** (B) Red Delicious
   (C) *Starking Delicious* (D) Richared Delicious

17. The Yellow stem borer of paddy belong to family
   (A) Crambidae (B) **Pyralidae**
   (C) Noctuidae (D) Pyraustidae

18. The Leaf folder borer of paddy belong to family
   (A) Crambidae (B) Pyralidae
   (C) Noctuidae (D) **Pyraustidae**

19. The Rice case worm belong to family
   (A) **Nymphalidae** (B) Pyralidae
   (C) Noctuidae (D) Pyraustidae

20. The Rice gall midge belong to family
    (A) Nymphalidae (B) Pyralidae
    (C) **Ceccidomyidae** (D) Pyraustidae

21. Which one of the following is a monophagous pest
    (A) *Orseolia oryzae* (B) ***Scirpophaga incertulas***
    (C) *Cnaphalocrocis medinalis* (D) All the above

22. Which one of the following is a monophagous pest
    (A) *Orseolia oryzae* (B) ***Scirpophaga incertulas***
    (C) *Cnaphalocrocis medinalis* (D) All the above

23. White ears and dead hearts are typical symptoms of infestation of
    (A) *Orseolia oryzae* (B) ***Scirpophaga incertulas***
    (C) *Cnaphalocrocis medinalis* (D) All the above

24. The scientific name of Rice gall midge is
    (A) ***Orseolia oryzae*** (B) *Scirpophaga incertulas*
    (C) *Cnaphalocrocis medinalis* (D) *Dicladispa armigera*

25. The scientific name of Rice hispa is
    (A) *Orseolia oryzae* (B) *Scirpophaga incertulas*
    (C) *Cnaphalocrocis medinalis* (D) ***Dicladispa armigera***

26. The Silver shoot/Onion leaf in rice is caused by
    (A) ***Orseolia oryzae*** (B) *Scirpophaga incertulas*
    (C) *Cnaphalocrocis medinalis* (D) *Dicladispa armigera*

27. The Clipping of leaf tip at transplanting is recommended for the control of
    (A) *Orseolia oryzae* (B) ***Scirpophaga incertulas***
    (C) *Cnaphalocrocis medinalis* (D) *Dicladispa armigera*

28. The Hopper burn is the typical symptom of
    (A) **Brown Plant Hopper** (B) Green Leaf hopper
    (C) White backed plant hopper (D) Zig zag leaf hopper

29. The scientific name of Brown plant hopper is
    (A) *Nephotettix viriscens* (B) *Sogatella furcifera*
    (C) ***Nilaparvata lugens*** (D) *Recilia dorsalis*

30. The vector of Orange leaf virus disease is
    (A) *Nephotettix viriscens* (B) *Sogatella furcifera*
    (C) *Nilaparvata lugens* (D) ***Recilia dorsalis***

31. The vector of mycoplasma Yellow dwarf disease is
   (A) *Nephotettix viriscens* (B) *Sogatella furcifera*
   (C) *Nilaparvata lugens* (D) ***Recilia dorsalis***

32. The scientific name of Green hopper is
   (A) ***Nephotettix viriscens*** (B) *Sogatella furcifera*
   (C) *Nilaparvata lugens* (D) *Recilia dorsalis*

33. The Grassy stunt virus is transmitted by
   (A) *Nephotettix viriscens* (B) *Sogatella furcifera*
   (C) ***Nilaparvata lugens*** (D) *Recilia dorsalis*

34. The Tungro virus is transmitted by
   (A) ***Nephotettix viriscens*** (B) *Sogatella furcifera*
   (C) *Nilaparvata lugens* (D) *Recilia dorsalis*

35. Planting with alley formation of 25 cm at intervals of 2mts is recommended against
   (A) *Nephotettix viriscens* (B) *Sogatella furcifera*
   (C) ***Nilaparvata lugens*** (D) *Recilia dorsalis*

36. Alternate wetting and drying is recommended for the control of
   (A) *Nephotettix viriscens* (B) *Sogatella furcifera*
   (C) *Nilaparvata lugens* (D) **All of the above**

37. *Cyrtorrhinus lividipennis* is a predator of
   (A) *Nephotettix viriscens* (B) *Sogatella furcifera*
   (C) ***Nilaparvata lugens*** (D) *Recilia dorsalis*

38. The release rate of *Cyrtorrhinus lividipennis* for BPH control is
   (A) 25-50 egg/m^2 (B) **50-75 egg/m^2**
   (C) 75-100 egg/m^2 (D) 100-150 egg/m^2

39. The larva fold leaf and feed with in causing longitudinal white lines is
   (A) *Orseolia oryzae* (B) *Scirpophaga incertulas*
   (C) ***Cnaphalocrocis medinalis*** (D) *Dicladispa armigera*

40. The unpleasant odour in the field and presence of individual chaffy/white ears is the symptom of
   (A) ***Leptocorisa acuta*** (B) *Scirpophaga incertulas*
   (C) *Cnaphalocrocis medinalis* (D) *Dicladispa armigera*

41. The *Dicladispa armigera* can be identified by
    (A) **Short pines on the body** (B) Hairs on the body
    (C) Both A and B (D) None of the above

42. The presence of white parallel streaks along the leaf is axis indicate the infestation of
    (A) *Leptocorisa acuta* (B) *Scirpophaga incertulas*
    (C) *Cnaphalocrocis medinalis* (D) ***Dicladispa armigera***

43. The marginal rolling of leaves in rice indicate the infestation of
    (A) *Leptocorisa acuta* (B) *Scirpophaga incertulas*
    (C) ***Stenchaeothrips biformis*** (D) *Dicladispa armigera*

44. The varieties showing resistance against stem borer
    (A) Ratna (B) Sasyashree
    (C) Vikas (D) **All of the above**

45. The bund trimming is recommended for the control of
    (A) *Leptocorisa acuta* (B) *Scirpophaga incertulas*
    (C) ***Hierpglyphus banian*** (D) *Dicladispa armigera*

46. The bund trimming is recommended for the control of
    (A) *Leptocorisa acuta* (B) *Scirpophaga incertulas*
    (C) ***Hierpglyphus banian*** (D) *Dicladispa armigera*

47. The insect-pests causing partial/complete white ear in paddy is
    (A) *Leptocorisa acuta* (B) *Scirpophaga incertulas*
    (C) *Hieropglyphus banian* (D) **All of the above**

48. Which of the following is not a pest of wheat
    (A) *Macrosiphum miscanthi* (B) *Helicoverpa armigera*
    (C) *Tanymecus indicus* (D) ***Chilo partellus***

49. The scientific name of Ghujhia weevil is
    (A) *Macrosiphum miscanthi* (B) *Helicoverpa armigera*
    (C) ***Tanymecus indicus*** (D) *Chilopartellus*

50. The scientific name of Wheat stem borer is
    (A) *Macrosiphum miscanthi* (B) *Helicoverpa armigera*
    (C) *Tanymecus indicus* (D) ***Sesamia inferens***

51. The armyworm is a major pest of
    (A) Sorghum (B) **Wheat**
    (C) Paddy (D) Bajra

52. The Molya disease in wheat is caused by

(A) ***Heterodera avenae*** (B) *Meloidogyne incognita*

(C) *Anguina tritici* (D) *Pratylenchus sp.*

53. The Molya disease in wheat is caused by

(A) Mite (B) Insect

(C) **Nematode** (D) Bacteria

54. The Ear-cockle or Mamni disease in wheat is caused by

(A) *Heterodera avenae* (B) *Meloidogyne incognita*

(C) ***Anguina tritici*** (D) *Pratylenchus sp.*

55. The Ear-cockle or Mamni disease of wheat is caused by*Anguina tritici* in association with

(A) ***Clavibacter tritici*** (B) *Pseudomonas flouorescens*

(C) *Xanthomonas tritici* (D) None of the above

56. The Ear-cockle or Mamni disease of wheat can be easily controlled by

(A) Using certified seeds (B) Seed floatation method

(C) **Both A and B** (D) None of the above

57. The formation of dead heart in wheat is the indication of infestation of

(A) *Macrosiphum miscanthi* (B) *Helicoverpa armigera*

(C) *Tanymecus indicus* (D) ***Sesamia inferens***

58. The damage of wheat leaf in semi circular fashion and partial root damage indicate the infestation of

(A) *Macrosiphum miscanthi* (B) ***Odontotermes obesus***

(C) *Tanymecus indicus* (D) *Sesamia inferens*

59. The damage of termites is more in

(A) Waterlogged areas (B) ***Dry areas***

(C) Irrigated areas (D) All of the above

60. The damage of wheat leaf in semi circular fashion and partial root damage indicate the infestation of

(A) *Macrosiphum miscanthi* (B) ***Odontotermes obesus***

(C) *Tanymecus indicus* (D) *Sesamia inferens*

61. Formation of dead hearts; reddening of midrib and pin hole damage on whorl of newly opened leaves in sorghum is caused by

(A) *Atherigona soccata* (B) ***Chilo partellus***

(C) *Calocoris angustatus* (D) *Contarinia sorghicola*

62. Red ooze when developing seeds in the spikelets are pressed in sorghum indicates the infestation of

(A) *Atherigona soccata* (B) *Chilo partellus*

(C) *Calocoris angustatus* (D) ***Contarinia sorghicola***

63. The scientific name of sorghum shoot bug is

(A) *Atherigona soccata* (B) ***Peregrinusmaidis***

(C) *Calocoris angustatus* (D) *Contarinia sorghicola*

64. The sorghum shoot bug belong to family

(A) Muscidae (B) Ceccidomyidae

(C) Miridae (D) **Delphacidae**

65. The *Atherigona soccata* belong to family

(A) **Muscidae** (B) Ceccidomyidae

(C) Miridae (D) Delphacidae

66. The *Calocoris angustatus* belong to family

(A) Muscidae (B) Ceccidomyidae

(C) **Miridae** (D) Delphacidae

67. The scientific name of maize stem borer is

(A) *Amsacta moorei* (B) ***Chilo partellus***

(C) *Agrotis ipsilon* (D) *Sesamia inferens*

68. The scientific name of European Corn borer is

(A) ***Ostrinia nubilalis*** (B) *Chilo partellus*

(C) *Agrotis ipsilon* (D) *Sesamia inferens*

69. The whorl feeding of larva resulting in rows of oblong holes is an indicator of

(A) *Amsacta moorei* (B) *Chilo partellus*

(C) *Agrotis ipsilon* (D) ***Sesamia inferens***

70. The whorl application of carbofuran @ 5g/whorl is recommended against

(A) *Amsacta moorei* (B) *Chilo partellus*

(C) *Agrotis ipsilon* (D) ***Sesamia inferens***

71. The Dead hearts and presence of pinhole damage is an indicator of

(A) *Amsacta moorei* (B) ***Chilo partellus***

(C) *Agrotis ipsilon* (D) *Sesamia inferens*

72. The whorl feeding of larva resulting in rows of oblong holes is an indicator of

(A) *Amsacta moorei* (B) *Chilo partellus*

(C) *Agrotis ipsilon* (D) ***Sesamia inferens***

73. The scientific name of groundnut leaf miner is
(A) *Amsacta moorei* (B) *Empoasca kerri*
(C) ***Aproaerma modicella*** (D) *Frankliniella schultzei*

74. The presence of white blotches, webbing and drying of terminal leaflet in groundnut are indictor of
(A) *Amsacta moorei* (B) *Empoasca kerri*
(C) ***Aproaerma modicella*** (D) *Frankliniella schultzei*

75. The *Aproaerma modicella* belong to family
(A) **Gelechiidae** (B) Noctuidae
(C) Cicadellidae (D) Thenthridinidae

76. The *Aproaerma modicella* pupates in
(A) Leaf (B) Soil
(C) Debris (D) **Webbing**

77. The downward curling of leaves and "V" shaped yellowing at the tip in groundnut is an indicator of
(A) *Amsacta moorei* (B) *Empoasca devestans*
(C) ***Empoasca kerri*** (D) *Frankliniella schultzei*

78. The *Empoasca kerri* lay eggs
(A) On the leaves (B) **Inserted in the leaf tissue**
(C) Inserted in the stem (D) On the flowers

79. The vector of Peanut stripe virus diseases is
(A) ***Aphis craccivora*** (B) *Empoasca kerri*
(C) *Aproaerma modicella* (D) *Frankliniella schultzei*

80. The vector of Peanut bud Necrosis virus diseases is
(A) *Aphis craccivora* (B) *Empoasca kerri*
(C) *Aproaerma modicella* (D) ***Frankliniella schultzei***

81. The scientific name of the Bihar hairy caterpillar is
(A) ***Spilosoma obliqua*** (B) *Amsacta moorei*
(C) *Aproaerma modicella* (D) *Frankliniella schultzei*

82. The Bihar hairy caterpillar pupates in
(A) On the leaves (B) **Soil**
(C) Inside the stem (D) All of the above

83. The scientific name of the Red Hairy caterpillar is

(A) *Spilosoma obliqua* (B) ***Amsacta moorei***

(C) *Aproaerma modicella* (D) *Frankliniella schultzei*

84. Which is a polyphagous pest

(A) *Spilosoma obliqua* (B) *Spodoptera litura*

(C) *Helicoverpa armigera* (D) **All of the above**

85. *Athalia lugens proxima* is a serious pest of

(A) **Mustard** (B) Astor

(C) Groundnut (D) Soybean

86. *Athalia lugens proxima* belong to family

(A) Pentatomidae (B) Eulophidae

(C) **Tenthridinidae** (D) Pterophoridae

87. Shot hole effect is the typical damage symptom of

(A) ***Athalia lugens proxima*** (B) *Spodoptera litura*

(C) *Myllocerus maculosus* (D) *Pieris brassicae*

88. The scientific name of castor capsule borer is

(A) ***Dichocrocis punctiferalis*** (B) *Spodoptera litura*

(C) *Parasa lepida* (D) *Pieris brassicae*

89. The scientific name of castor Slug is

(A) *Dichocrocis punctiferalis* (B) *Spodoptera litura*

(C) ***Parasa lepida*** (D) *Pieris brassicae*

90. The castor Slug, *Parasa lepida* belong to family

(A) **Limacodidae** (B) Pyralidae

(C) Sphingidae (D) Noctuidae

91. *Antigastra catalaunalis* is a serious pest of

(A) Castor (B) **Sesame**

(C) Safflower (D) Sunflower

92. The scientific name of castor whitefly is

(A) *Bemesia tabaci* (B) ***Trialeurodes ricini***

(C) *Trialeurodes dispersus* (D) None of the above

93. The castor capsule borer, *Conogenothes puctiferalis* belong to family

(A) **Limacodidae** (B) Pyralidae

(C) Sphingidae (D) Noctuidae

94. *Dasyneura sesami* , linseed gall fly belong to family

(A) **Ceccidomyidae** (B) Agromyzidae
(C) Muscidae (D) None of the above

95. *Perigaea capensis* is a serious pest of

(A) Castor (B) Sesame
(C) **Safflower** (D) Sunflower

96. *Acnthiophilus helianthi,* safflower bud fly belong to family

(A) Ceccidomyidae (B) Agromyzidae
(C) Muscidae (D) **Tephritidae**

97. *Antigastra catalaunalis* is a serious pest of

(A) Castor (B) **Sesame**
(C) Safflower (D) Sunflower

98. Mustard sawfly lay eggs

(A) **Singly inserted near leaf margin** (B) Singly on leaves
(C) In groups on leaves (D) In the soil

99. Mustard sawfly is characterized by having

(A) 5 pair of prolegs (B) 6 pair of prolegs
(C) 7 pair of prolegs (D) **8 pair of prolegs**

100. *Scirpophaga incertulas* lay eggs

(A) **Egg mass at leaf tip** (B) Egg batches on leaf
(C) Singly on leaf (D) Singly in the stem

101. *Scirpophaga incertulas* hibernates in

(A) Leaves (B) Soil
(C) Debris (D) **At the base of the stem**

102. *Dasyneura sesami* , linseed gall fly belong to family

(A) **Ceccidomyidae** (B) Agromyzidae
(C) Muscidae (D) None of the above

103. *Acanthiophilus helianthi,* Safflower bud fly pupates in

(A) Leaves (B) **Flower buds**
(C) Soil (D) Debris

104. *Asphondylia sesami,* Linseed gall fly pupates in

(A) Leaves (B) Flower buds
(C) **Deformed capsules** (D) Debris

105. Flowers and young capsules with gall like swelling in Linseed is the typical symptom of attack of which insect pest

(A) ***Asphondylia sesami*** (B) *Thysanoplusia orichalcea*

(C) *Antigastra catalaunalis* (D) *Uroleucon carthami*

106. The varieties shown resistance against sorghum shoot fly is

(A) CHS-7 (B) CHS-8

(C) IS-5566 (D) **All of the above**

107. Flowers and young capsules with gall like swelling in Linseed is the typical symptom of attack of which insect pest

(A) ***Asphondylia sesami*** (B) *Thysanoplusia orichalcea*

(C) *Antigastra catalaunalis* (D) *Uroleucon carthami*

108. *Helicoverpa armigera* lay eggs

(A) **Singly** (B) In groups

(C) In egg mass (D) In batches

109. Which larval instar of *Helicoverpa armigera* bore the pods

(A) 2nd (B) **3rd**

(C) 4th (D) 5th

110. *Helicoverpa armigera* pupates in

(A) Leaves (B) **Soil**

(C) Debris (D) All of the above

111. A single *Helicoverpa armigera* feed on how many pod before reaching maturity

(A) 10-20 (B) **30-40**

(C) 50-60 (D) 70-80

112. *Melanoagromyza obtuse* is a major pest of

(A) Gram (B) **Pigeon pea**

(C) Green gram (D) Cowpea

113. The scientific name of Legume pod borer/spotted caterpillar is

(A) ***Maruca vitrata*** (B) *Lampide boeticus*

(C) *Autographa nigrisigna* (D) *Melanogromyza obtuse*

114. Which of the following is a monophagous pest

(A) *Melanogromyza obtuse* (B) *Exlastis atomosa*

(C) **Both A and B** (D) None of the above

115. The Tur pod fly, *Melanoagromyza obtuse* lay eggs
   (A) **Singly inserted in the pod shell** (B) Singly on pods
   (C) Singly on the flowers (D) All of the above
116. The *Maruca vitrata* lay eggs
   (A) Singly on the pod (B) Singly on the flowers
   (C) Singly on the buds (D) **All of the above**
117. The *Lampides boeticus* belong to family
   (A) Pyraustidae (B) Crambidae
   (C) Pyralidae (D) **Lycaenidae**
118. The Plume moth, *Exlastis atomosa* belong to family
   (A) Pyraustidae (B) Crambidae
   (C) **Pterophoridae** (D) Lycaenidae
119. The lentil pod borer, *Etiella zinckiniella* pupates in
   (A) On the pods (B) On the flowers
   (C) **Soil** (D) All of the above
120. The lentil pod borer, *Etiella zinckiniella* pupates in
   (A) On the pods (B) On the flowers
   (C) **Soil** (D) All of the above
121. The Leaf binder, *Grapholita critica* belong to family
   (A) **Totricidae** (B) Crambidae
   (C) Pterophoridae (D) Lycaenidae
122. The vector of mungbean yellow mosaic virus is
   (A) Bean thrips (B) **hitefly**
   (C) Bean aphid (D) Bean Jassid
123. The vector of Pea leaf roll virus is
   (A) Bean thrips (B) Whitefly
   (C) **Bean aphid** (D) Bean Jassid
124. The bean aphid, *Aphis craccivora* is a vector of
   (A) Cucumber mosaic virus (B) Pea leaf roll virus
   (C) Tobacco streak virus (D) **All of the above**
125. The Mites are mostly dispersed through
   (A) **Wind** (B) Seed
   (C) Plant (D) Water

126. The Presence of 'Hot spot" indicate the attack of
(A) Thrips (B) Viruses
(C) Jassids (D) **Mites**

127. *Chromatomyia horticola,* Pea leaf miner belong to family
(A) Ceccidomyidae (B) **Agromyzidae**
(C) Muscidae (D) Tephritidae

128. The larva of blister beetle is known as
(A) Protopod (B) **Triungulins**
(C) Caterpillar (D) Maggot

129. The blister beetle pupate in
(A) **Soil** (B) Seed
(C) Leaves (D) All of the above

130. Drooping of the tender leaves and yellowing of pea indicates the attack of
(A) *Chromatomyia horticola* (B) *Lampides boeticus*
(C) ***Ophiomyia phaseoli*** (D) None of the above

131. The Gram cut worm, *Agrotis ipsilon* cut the plant at height of
(A) **5cm from base** (B) 10cm from base
(C) 15cm from base (D) 2.5cm from the base

132. Which of the following is both field and storage pest
(A) *Chromatomyia horticola* (B) *Lampides boeticus*
(C) *Ophiomyia phaseoli* (D) ***Callosobruchus chinensis***

133. The scientific name of the spotted bean bruchid is
(A) *Callosobruchus chinensis* (B) ***Callosobruchus maculatus***
(C) *Callosobruchus phaseoli* (D) *Callosobruchus analis*

134. The scientific name of the Lentil bruchid/Dhora is
(A) *Callosobruchus chinensis* (B) *Callosobruchus maculatus*
(C) ***Callosobruchus phaseoli*** (D) *Callosobruchus analis*

135. The scientific name of the Moong bruchid/dhora is
(A) *Callosobruchus chinensis* (B) *Callosobruchus maculatus*
(C) *Callosobruchus phaseoli* (D) ***Callosobruchus analis***

136. The scientific name of the Pea bruchid/dhora is
(A) ***Callosobruchus affinis*** (B) *Callosobruchus maculatus*
(C) *Callosobruchus phaseoli* (D) *Callosobruchus analis*

137. The scientific name of the Pea bruchid/dhora is

(A) ***Callosobruchus affinis*** (B) *Callosobruchus maculatus*

(C) *Callosobruchus phaseoli* (D) *Callosobruchus analis*

138. The application of which pesticide group lead to resurgence of *Helicoverpa armigera*

(A) Organophosphates (B) **Synthetic pyrethroids**

(C) Spinosad (D) Carbamates

139. The application of which pesticide group lead to whitefly resurgence

(A) Organophosphates (B) **Synthetic pyrethroids**

(C) Spinosad (D) Carbamates

140. The application of which pesticide group lead to resurgence of mealybugs

(A) Organophosphates (B) Synthetic pyrethroids

(C) **Spinosad** (D) Carbamates

141. The scientific name of sugarcane shoot borer is

(A) ***Chilo infuscatellus*** (B) *Scirpophaga nivella*

(C) *Chilo auricilius* (D) *Acigona steniella*

142. The scientific name of sugarcane top shoot borer is

(A) *Chilo infuscatellus* (B) ***Scirpophaga nivella***

(C) *Chilo auricilius* (D) *Acigona steniella*

143. The scientifc name of Gurdaspur borer is

(A) *Chilo infuscatellus* (B) *Scirpophaga nivella*

(C) *Chilo auricilius* (D) ***Acigona steniella***

144. The scientific name of sugarcane shoot borer is

(A) ***Chilo infuscatellus*** (B) *Scirpophaga nivella*

(C) *Chilo auricilius* (D) *Acigona steniella*

145. The scientific name of sugarcane stalk borer is

(A) *Chilo infuscatellus* (B) *Scirpophaga nivella*

(C) ***Chilo auricilius*** (D) *Acigona steniella*

146. The scientific name of sugarcane Plassey borer is

(A) *Chilo saccharophagus indicus* (B) *Emmalocera depresella*

(C) *Raphimetopus albutella* (D) ***Chilo tumidicostalis***

147. The scientific name of sugarcane Internode borer is

(A) ***Chilo saccharophagus indicus*** (B) *Emmalocera depresella*

(C) *Raphimetopus albutella* (D) *Chilo tumidicostalis*

148. The scientific name of sugarcane Green borer is
   (A) *Chilo saccharophagus indicus*   (B) *Emmalocera depresella*
   (C) ***Raphimetopus albutella***   (D) *Chilo tumidicostalis*

149. The scientific name of sugarcane Root borer is
   (A) *Chilo saccharophagus indicus*   (B) *Emmalocera depresella*
   (C) *Raphimetopus albutella*   (D) ***Chilo tumidicostalis***

150. The drying of central shoots resulting in bunchy top and stunted growth of sugarcane is due to
   (A) *Chilo infuscatellus*   (B) ***Scirpophaga nivella***
   (C) *Chilo auricilius*   (D) *Acigona steniella*

151. The presence of make semicircular tunnels and withering of central shoots resulting in dead hearts in sugarcane is due to
   (A) *Chilos accharophagus indicus*   (B) ***Emmalocera depresella***
   (C) *Raphimetopus albutella*   (D) *Chilo tumidicostalis*

152. The parasitoid of sugarcane pyrilla is
   (A) ***Epiricania melanoleuca***   (B) *Trichogramma chilonis*
   (C) *Bracon hebetor*   (D) *Telenomus remes*

153. The release rate of *Epiricania melanoleuca* to control sugarcane pyrilla is
   (A) 4000-5000 cocoons/ha   (B) 4-5 lakh eggs/ha
   (C) **Both A and B**   (D) None of the above

154. Gappy appearance of seedlings and sugarcane setts filed with soil indicate the infestation of
   (A) **Termites**   (B) Mealybugs
   (C) Root borer   (D) Mites

155. Clipping of leaves in sugarcane help in reducing infestation of
   (A) Top borer   (B) Gurdaspur borer
   (C) Pyrilla   (D) **All of the above**

156. Intercropping of Dhaincha in sugarcane help in reducing the infestation of
   (A) Top borer   (B) Gurdaspur borer
   (C) Pyrilla   (D) **Shoot borer**

157. The scientific name of sugarcane whitefly is
   (A) *Bemesia tabaci*   (B) ***Trialeurodes ricini***
   (C) *Trialeurodes dispersus*   (D) *Aleurolobus barodensis*

158. The Mite infesting Sugarcane is

(A) *Schizotetranychus andropogoni* (B) *Oligonychus indicus*

(C) *Aceria sacchari* (D) **All of the above**

159. The release rate of *Melanaspis glomerata* to control s ugarcane scale is

(A) 500 adults or grubs /ha (B) 1000 adults or grubs /ha

(C) **1500 adults or grubs /ha** (D) 2000 adults or grubs /ha

160. The release rate of *Isotima javensis* to control sugarcane top borer is

(A) 50 adults /ha thrice (B) **115 adults /ha thrice**

(C) 250 adults /ha thrice (D) 300adults /ha thrice

161. The Sunhemp stem borer, *Enarmonia pesudonectis* belong to family

(A) Lamiidae (B) **Tortricidae**

(C) Arctiidae (D) Miridae

162.The total area under Bt cotton on India is

(A) 9.60 mha (B) **11.60mha**

(C) 14.60mha (D) 17.60 mha

163. The state having largest area under Bt cotton in India is

(A) Gujrat (B) Punjab

(C) **Maharashtra** (D) Karnataka

164. The First indigenous variety having Bt gene in India is

(A) NHH-44 (B) **Bikaneri Nerma**

(C) MLS-9142 (D) None of the above

165. The Major company supplying Bt cotton in India in India is

(A) Syngenta (B) **Mahyco/Monsanto**

(C) Bayer (D) Rallis India

166.The First Bt cotton varieties approved for cultivation in India is

(A) MECH-12 (B) MECH-184

(C) MECH-162 (D) **All of the above**

167. The Sunhemp capsid, *Ragmus importunitas* belong to family

(A) Lamiidae (B) Tortricidae

(C) Arctiidae (D) **Miridae**

168. The cotton leaves rolling like a trumpet/funnel is the damage symptom of

(A) *Amrasca bigutulla biguttula* (B) ***Sylepta derogata***

(C) *Phenacoccus solenopsis* (D) *Pempherelus affinis*

169. The outbreak of which insect on cotton was reported in 2015 from Punjab

(A) *Amrasca bigutulla biguttula* (B) *Sylepta derogata*

(C) ***Bemesia tabaci*** (D) *Pempherelus affinis*

170. The Larva head bore in to the boll, leaving rest of the body outside is

(A) *Erias vittella* (B) ***Helicoverpa armigera***

(C) *Erias insulana* (D) *Pectinophora gossypiella*

171. The large and circular entry hole at the base of the boll is due to

(A) *Erias vittella* (B) ***Helicoverpa armigera***

(C) *Erias insulana* (D) *Pectinophora gossypiella*

172. The Holes on damaged boll plugged with excreta leading to rotting of lint

(A) *Erias vittella* (B) *Helicoverpa armigera*

(C) ***Erias insulana*** (D) *Pectinophora gossypiella*

173. The Double seed is the characteristic of

(A) *Erias vittella* (B) *Helicoverpa armigera*

(C) *Erias insulana* (D) ***Pectinophora gossypiella***

174. The Holes on damaged boll plugged with excreta leading to rotting of lint

(A) *Erias vittella* (B) *Helicoverpa armigera*

(C) ***Erias insulana*** (D) *Pectinophora gossypiella*

175. The vector of cotton leaf curl virus is

(A) ***Bemesia tabaci*** (B) *Trialeurodes ricini*

(C) *Trialeurodes dispersus* (D) *Aleurolobus barodensis*

176. The vector of Vein clearing disease of okra is

(A) ***Bemesia tabaci*** (B) *Trialeurodes ricini*

(C) *Trialeurodes dispersus* (D) *Aleurolobus barodensis*

177. The vector of Sesamum leaf curl virus is

(A) ***Bemesia tabaci*** (B) *Trialeurodes ricini*

(C) *Trialeurodes dispersus* (D) *Aleurolobus barodensis*

178. The scientific name of cotton semilooper is

(A) *Anomis flava* (B) ***Tarache notabilis***

(C) *Phycita infusella* (D) *Pempherelus affinis*

179. The scientific name of Green semilooper is

(A) ***Anomis flava*** (B) *Tarache notabilis*

(C) *Phycita infusella* (D) *Pempherelus affinis*

180. The *Pectinophora gossypiella* belong to family

(A) Noctuidae (B) **Gelechidae**

(C) Pyralidae (D) Pyraustidae

181. The *Pectinophora gossypiella pupate* in

(A) Leaves (B) **Boll**

(C) Plant (D) Debris

182. The ETL value for pink bollworm in cotton is

(A) 1-5% boll infestation (B) **5-10% fruit infestation**

(C) 10-15% boll infestation (D) >15% shoot infestation

183. The ETL value for American bollworm in cotton is

(A) 1-5% boll infestation (B) **5-10% fruit infestation**

(C) 10-15% boll infestation (D) >15% shoot infestation

184. The ETL value for whitefly in cotton is

(A) 2-5 adults/leaf (B) **8-10 adults/leaf**

(C) 10-15 adults/leaf (D) >15adults/leaf

185. Trap crop recommended for Bollworms in cotton is

(A) **Okra** (B) Castor

(C) Marigold (D) Cucumber

186. Okra leaves type cultivars in Cotton are resistant to

(A) American bollworm (B) Pink bollworm

(C) Whiteflies (D) **All of the above**

187. Trap crop recommended for reducing thrips infestation in cotton is

(A) Garlic (B) Onion

(C) **Both A and B** (D) Castor

188. The tolerant varieties against PBW is

(A) JHK-1 (B) Khandwa -2

(C) Abhadita (D) **All of the above**

189. Yellow sticky traps are recommended against

(A) Aphids (B) Whiteflies

(C) Jassids (D) **All of the above**

190. The scientific name of Jute stem girdler is

(A) *Anomis sabulifera* (B) ***Nupserha bicolorpostbrunnea***

(C) *Apion corchori* (D) *Utethesia pulchella*

191.The Jute stem girdler, *Nupserha bicolor* belong to family

(A) **Lamiidae** (B) Tortricidae

(C) Arctiidae (D) Miridae

192. The scientific name of Jute stem weevil is

(A) *Anomis sabulifera* (B) *Nupserha bicolor*

(C) ***Apion corchori*** (D) *Utethesia pulchella*

193. The scientific name of Jute semilooper is

(A) ***Anomis sabulifera*** (B) *Nupserha bicolor*

(C) *Apion corchori* (D) *Utethesia pulchella*

194. The scientific name of Sunhemp hairy caterpillar is

(A) *Anomis sabulifera* (B) *Nupserha bicolor*

(C) *Apion corchori* (D) ***Utethesia pulchella***

195. The scientific name of Sunhemp stem borer is

(A) *Ragmus importunitas* (B) *Nupserha bicolor*

(C) ***Enarmonia pesudonectis*** (D) *Utethesia pulchella*

196. The scientific name of Sunhemp capsid is

(A) ***Ragmus importunitas*** (B) *Nupserha bicolor*

(C) *Enarmonia pesudonectis* (D) *Utethesia pulchella*

197. The scientific name of Sunhemp hairy caterpillar is

(A) *Ragmus importunitas* (B) *Nupserha bicolor*

(C) *Enarmonia pesudonectis* (D) ***Utethesia pulchella***

198. The Sunhemp hairy caterpillar, *Utethesia pulchella* belong to family

(A) Lamiidae (B) Tortricidae

(C) **Arctiidae** (D) Miridae

Unit-10

Insect Pests of Horticultural Crops

Unit-10

Insect Pests of Horticultural Crops

Q 1. Match the following

1. The author of "*Insects and Fruits*" is

(A) Atwal AS & Dhaliwal GS (B) MRGK Nair

(C) **Butani DK & Jotwani MG.** (D) AR Horowitz and I Ishaya

2. The author of "Insect and mites of crops in India" is

(A) Atwal AS & Dhaliwal GS (B) **MRGK Nair**

(C) Butani DK & Jotwani MG. (D) AR Horowitz and I Ishaya

3. The author of "*Insects and Vegetables*" is

(A) Atwal AS & Dhaliwal GS (B) MRGK Nair

(C) **Butani DK & Jotwani MG.** (D) AR Horowitz and I Ishaya

4. The author of "*Agricultural Pests of South Asia and their Management*" is

(A) **Atwal AS & Dhaliwal GS** (B) MRGK Nair

(C) Butani DK & Jotwani MG. (D) AR Horowitz and I Ishaya

5. The Hadda beetle, *Henosepilachna vigintioctpuctata* has spots on each eltyra

(A) <7 (B) >7

(C) 7 (D) **7-14**

6. The Hadda beetle, *Henosepilachna dodecastigma* has spots on each eltyra

(A) <7 (B) **6**

(C) 7 (D) 7-14

7. The overwintering stage of Hadda beetle is

(A) Larval (B) **Adult**

(C) Pupal (D) Egg

8. The eggs of Hadda beetle are

(A) **Yellow Cigar shape** (B) Yellow Oval shape

(C) White oval shape (D) White Oval shape

9. The varieties tolerant to Hadda beetle is

(A) Arka Shirish (B) Hissar

(C) Shankar Vijai (D) **All of the above**

10. The damaging stage of Hadda beetle is

(A) Larval stage (B) Adult stage

(C) **Both Larval and Adult stage** (D) None of the above

11. The Scientific name of Brinjal Lacewing bug is

(A) *Euzophera perticella* (B) ***Urentius sentis***

(C) *Leucinoides orbonalis* (D) None of the above

12. The characteristic mottled appearance of Brinjal leaves is due to

(A) *Euzophera perticella* (B) ***Urentius sentis***

(C) *Leucinoides orbonalis* (D) None of the above

13. The Scientific name of Brinjal stem borer is

(A) ***Euzophera perticella*** (B) *Urentius sentis*

(C) *Leucinoides orbonalis* (D) None of the above

14. The Scientific name of Brinjal shoot and fruit borer is

(A) *Euzophera perticella* (B) *Urentius sentis*

(C) ***Leucinoides orbonalis*** (D) None of the above

15. The Brinjal shoot and fruit borer belong to family

(A) Tingidae (B) **Pyraustidae**

(C) Pyralidae (D) None of the above

16. The hibernating stage of Brinjal shoot and fruit borer is

(A) **Larva** (B) Pupa

(C) Adult (D) None of the above

17. The Brinjal shoot and fruit borer lay eggs

(A) Egg mass (B) **Singly**

(C) Ingroups (D) None of the above

18. The Brinjal shoot and fruit borer Pupate in

(A) Fruits (B) **Fallen leaves**

(C) Soil (D) None of the above

19. The drooping of terminal shoots and presence of bore fruits is a typical damage symptom of

(A) *Euzophera perticella* (B) *Urentiussentis*

(C) ***Leucinoides orbonalis*** (D) None of the above

20. In general, one larva of Brinjal shoot and fruit borer damage

(A) 1-2 fruits (B) 2-4 fruits

(C) **4-6 fruits** (D) 8-10 fruits

21. The manag ement strategy against Brinjal shoot and fruit borer is

(A) Clipping of infested shoots (B) Pheromone traps

(C) Early transplanting (D) **All of the above**

22. The cultivars having tolerance against Brinjal shoot and fruit borer is

(A) Pusa Purple Cluster (B) Pant Samrat

(C) Punjab Neelum (D) **All of the above**

23. The Wota pheromone traps are used against Brinjal shoot and fruit borer as

(A) BSFB is a strong flier (B) **BSFB is a weak flier**

(C) Both A and B (D) None of the above

24. Exclusive stem feeder of brinjal is

(A) ***Euzophera perticella*** (B) *Urentius sentis*

(C) *Leucinoides orbonalis* (D) None of the above

25. The little leaf of brinjal is transmitted by

(A) *Euzophera perticella* (B) *Urentius sentis*

(C) *Leucinoides orbonalis* (D) ***Cestius phycitis***

26. The little leaf of brinjal is a

(A) Bacteria disease (B) Viral disease

(C) **Mycoplasma disease** (D) Protozoan disease

27. The scientific name of Tomato fruit borer is

(A) *Euzophera perticella* (B) *Urentius sentis*

(C) *Leucinoides orbonalis* (D) ***Helicoverpa armigera***

28. The *Helicoverpa armigera* lay eggs

(A) **Singly** (B) In groups

(C) Egg mass (D) **All of the above**

29. Which is true about *Helicoverpa armigera*

(A) Colour variation (B) Cannibalistic

(C) Polyphagous (D) **All of the above**

30. The trap crop recommended for *Helicoverpa armigera* management is

(A) **African Marigold** (B) Okra

(C) Black gram (D) Sorghum

31.The proportion of marigold : Tomato crop in trap cropping is

(A) 1:10 (B) **1:16**

(C) 1:20 (D) 1:6

32. The trap crop recommended for *Helicoverpa armigera* management is

(A) **Marigold** (B) Okra

(C) Black gram (D) Sorghum

33. The Trichogramma species recommended for *Helicoverpa armigera* management in tomato is

(A) *Trichogramma chilonis* (B) *Trichogramma brasiliensis*

(C) *Trichogramma embryophagum* (D) ***Trichogramma pretiosum***

34. The release rate of *Trichogramma pretiosum* for *Helicoverpa armigera* management is

(A) **50,000/ha; six release** (B) 1,00,000/ha; six release

(C) 1,50,000/ha; six release (D) 2,00,000/ha; six release

35. The Tomato whitefly belong to family

(A) **Aleurodidae** (B) Coccidae

(C) Miridae (D) Diaspididae

36. The Leaf curl virus in tomato is transmitted by

(A) ***Bemesia tabaci*** (B) *Trialeurodes ricini*

(C) *Trialeurodes dispersus* (D) *Aleurolobus barodensis*

37. The Murda disease in Chillies is caused by

(A) *Scirtothrips dorsalis* (B) *Trialeurodes ricini*

(C) ***Polyphagotarsonemus latus*** (D) *Aleurolobus barodensis*

38. The affected leaves are inverted boat; internodes become shorter in Chillies due to the infestation of

(A) *Scirtothrips dorsalis* (B) *Trialeurodes ricini*

(C) ***Polyphagotarsonemus latus*** (D) *Aleurolobus barodensis*

39. The affected leaves are boat shaped; internodes normal in Chillies due the infestation of

(A) ***Scirtothrips dorsalis*** (B) *Trialeurodes ricini*

(C) *Polyphagotarsonemus latus* (D) *Aleurolobus barodensis*

40. The vector of chillies leaf virus is

(A) *Scirtothrips dorsalis* (B) ***Bemesia tabaci***

(C) *Polyphagotarsonemus latus* (D) *Aleurolobus barodensis*

41. The affected leaves are boat shaped; internodes normal in Chillies due to the infestation of

(A) ***Scirtothrips dorsalis*** (B) *Trialeurodes ricini*
(C) *Polyphagotarsonemus latus* (D) *Aleurolobus barodensis*

42. The scientific name of brinjal red mite is

(A) *Tetranyychus urticae* (B) ***Tetranychus cinnabarinus***
(C) *Tetranychus neocaledonics* (D) *Aleurolobus barodensis*

43. The scientific name of Okra red mite is

(A) ***Tetranyychus urticae*** (B) *Tetranychus cinnabarinus*
(C) *Tetranychus neocaledonics* (D) *Aleurolobus barodensis*

44. The scientific name of Tomato red mite is

(A) *Tetranychus neocaledonics* (B) *Aceria lycopersici*
(C) **Both A and B** (D) *Tetranychus cinnabarinus*

45. The scientific name of Potato tuber moth is

(A) *Spodoptera litura* (B) ***Phthorimaea operculella***
(C) *Polyphagotarsonemus latus* (D) *Agrotis ipsilon*

46. The insect pest of potato which infest both in field and in storage is

(A) *Spodoptera litura* (B) ***Phthorimaea operculella***
(C) *Polyphagotarsonemus latus* (D) *Agrotis ipsilon*

47. The characteristic symptoms of PTM damage is

(A) Mines on the leaves (B) Bore hole on tender shoots
(C) **Both A and B** (D) Scratching of leaves

48. The scientific name of Potato cutworm is

(A) *Spodoptera litura* (B) *Phthorimaea operculella*
(C) *Polyphagotarsonemus latus* (D) ***Agrotis ipsilon***

49. The scientific name of Potato tuber moth is

(A) *Spodoptera litura* (B) ***Phthorimaea operculella***
(C) *Polyphagotarsonemus latus* (D) *Agrotis ipsilon*

50. The scientific name of Golden Cyst nematode is

(A) *Meloidogyne incognita* (B) *Heterodera avenae*
(C) ***Globodera rostochiensis*** (D) *Ditylenchus angustatus*

51. The nematode having quarantine significance in potato is

(A) *Meloidogyne incognita* (B) *Heterodera avenae*
(C) ***Globodera rostochiensis*** (D) *Ditylenchus angustatus*

52. The quarantine on *Globodera rostochiensis* is imposed in

(A) Jammu, J&K (B) Kufri, Shimla

(C) **Darjeeling, W.B** (D) Pant Nagar, Uttarakhand

53. The infective stage of *Globodera rostochiensis*

(A) IJ 1 (B) **IJ 2**

(C) IJ 3 (D) IJ 4

54. The scientific name of Red Pumpkin beetle is

(A) *Plusia orichalcea* (B) ***Aulacophora foveocollis***

(C) *Phyllotreta cruciferae* (D) None of the above

55. The family of Red Pumpkin beetle is

(A) **Chrysomelidae** (B) Cerambycidae

(C) Dynastidae (D) None of the above

56. The damage symptoms of Red Pumpkin beetle is

(A) Adult cut irregular bore holes on the leaves

(B) Grubs borer in the roots and fruits touching soil

(C) **Both A and B**

(D) None of the above

57. The damaging stage of Red Pumpkin beetle is

(A) Adults (B) Grubs

(C) **Both Adult and Grubs** (D) None of the above

58. The Red Pumpkin beetle lay eggs

(A) **In soil** (B) On leaves

(C) On fruits (D) All of the above

59. The Red Pumpkin beetle pupates

(A) **In soil** (B) On leaves

(C) On fruits (D) All of the above

60. The fruit fly infesting cucurbits is

(A) *Bactrocera dorsalis* (B) ***Bactrocera cucurbitae***

(C) *Carpomyia vesuviana* (D) None of the above

61. The Incidence of fruit fly on cucurbits is more in

(A) **In rainy season** (B) In Summer

(C) In Winter (D) Throughout the year

62. The fruit fly lay egg
    (A) **Inside the fruits** (B) On the fruits
    (C) On leaves (D) Inside the leaves

63. The typical damage symptoms of fruit fly attack is
    (A) Premature dropping of fruits
    (B) Oozing of brown resinous fluid from infested fruits
    (C) Fruits become distorted and malformed
    (D) **All of the above**

64. The fruit fly larvae pupates is
    (A) **In soil** (B) On leaves
    (C) On fruits (D) All of the above

65. The management strategy for fruit fly are
    (A) Frequent raking of soil (B) Use of fruit fly traps
    (C) Variation in sowing time (D) **All of the above**

66. The composition of bait spray required in one hectare for fruit fly control is
    (A) **Sugar (0.5kg) + Malathion (50ml)+ water (50L)**
    (B) Sugar (1.00kg) + Malathion (50ml)+ water (50L)
    (C) Sugar (1.00kg) + Malathion (100ml)+ water (50L)
    (D) All of the above

67. The ETL for brinjal shoot and fruit borer is
    (A) **1-5% shoot/fruit infestation** (B) 5-10% shoot/fruit infestation
    (C) 10-15% shoot/fruit infestation (D) None of the above

68. The ETL for Tomato fruit borer is
    (A) **1-5% fruit damage** (B) 5-10% fruit damage
    (C) 10-15% fruit damage (D) None of the above

69. The *Amrasca biguttula biguttula* is a serious pest of
    (A) Beans (B) **Okra**
    (C) Brinjal (D) Tomato

70. The Okra jassid, *Amrasca biguttula biguttula* lay eggs
    (A) On the leaves (B) **Inside the leaf tissue**
    (C) On the fruits (D) Inside the plants parts

71. The yellowing and curling of leaves along the margins is the damage symptom of
    (A) ***Amrasca biguttula biguttula*** (B) *Bemesia tabaci*
    (C) *Aceria lycopersici* (D) All of the above

72.The fruit borer infesting okra is

(A) *Erias insulana* (B) *Erias vittella*

(C) **Both A and B** (D) None of the above

73. The fruit borer of okra prevalent in dry areas is

(A) ***Erias insulana*** (B) *Erias vittella*

(C) Both A and B (D) None of the above

74. The fruit borer of okra prevalent in humid areas is

(A) *Erias insulana* (B) ***Erias vittella***

(C) Both A and B (D) None of the above

75. The cultivars resistance against okra jassid is

(A) Punjab -7 (B) Vaishali Madhu

(C) Sel-6 (D) **All of the above**

76. The cultivars resistance against okra fruit borer is

(A) Pusa Sawani (B) Vaishali Madhu

(C) Sel-6 (D) **All of the above**

77. The fruit borer of okra, *E. insulana* lay eggs

(A) **Singly on buds and flower** (B) Egg mass on buds and flower

(C) Singly on leaves (D) Egg mass on leaves

78. The Blister beetle is a serious pest of

(A) Okra (B) Maize

(C) **Both A and B** (D) None of the above

79. The yellow vein mosaic virus in okra is transmitted by

(A) *Scirtothrips dorsalis* (B) ***Bemesia tabaci***

(C) *Aphis gossypii* (D) All of the above

80. The thrips infesting onion is

(A) *Scirtothrips dorsalis* (B) *Trialeurodes dispersus*

(C) ***Thrips tabaci*** (D) None of the above

81. The Onion thrips, *Thrips tabaci lay* eggs

(A) On the leaves (B) **Inside the leaf tissue**

(C) On the fruits (D) Inside the plants parts

82. The Onion thrips, *Thrips tabaci pupate* in

(A) On the leaves (B) Inside the leaf

(C) On the fruits (D) **In the soil**

83. The Onion thrips, *Thrips tabaci* is a vector of
   (A) Peas Streak virus (B) Pineapple yellow spot virus
   (C) **Both A and B** (D) None of the above

84. The Onion fly belong to family
   (A) Tephritidae (B) Agromyzidae
   (C) Muscidae (D) **Anthomyiidae**

85. The Onion fly pupates in
   (A) Shoots (B) Roots
   (C) **Soil** (D) Fruit

86. The soft rot of onion is associated with
   (A) *Thrips tabaci* (B) ***Delia antiqua***
   (C) *Bemesia tabaci* (D) All of the above

87. The Pea leaf miner belong to family
   (A) Tephritidae (B) **Agromyzidae**
   (C) Muscidae (D) Anthomyiidae

88. The Pea leaf miner overwinter in
   (A) Egg stage (B) Larval stage
   (C) **Pupal stage** (D) Adult stage

89. The Pea stem fly belong to family
   (A) Tephritidae (B) **Agromyzidae**
   (C) Muscidae (D) Anthomyiidae

90. The scientific name of pea pod borer
   (A) *Helicoverpa armigera* (B) *Erias insulana*
   (C) *Pectinophora gossypiella* (D) ***Etiella zinckiniella***

91. The damaging stage of Singhara beetle is
   (A) Grub stage (B) Adult stage
   (C) **Both Grub and Adult stage** (D) None of the above

92. The Singhara beetle is a major pest of
   (A) **Water nuts** (B) Chestnuts
   (C) Cashewnut (D) None of the above

93. The sweet potato weevil is a major pest of
   (A) Field sweet potato (B) Stored sweet potato
   (C) **Both A and B** (D) None of the above

94. The scientific name of sweet potato weevil is

(A) *Omphisa anastomosalis* (B) ***Cylas formicarius***

(C) *Phthorimaea opercullela* (D) None of the above

95. The *Pieris brassciace* lay eggs

(A) Yellowish conical eggs singly (B) White conical eggs singly

(C) **Yellowish conical eggs in cluster** (D) White conical eggs clusters

96. The Diamond Backmothbelong to family

(A) **Plutellidae** (B) Pieridae

(C) Noctuidae (D) Crambidae

97. The Barrel shape cocoon is present in

(A) *Pieris brasscicae* (B) ***Plutella xylostella***

(C) *Crocidolomia binotalis* (D) *Hellula undalis*

98. The typical damage symptoms of DBM is

(A) Mining of leaves in early instars

(B) Larvae feed on leaves leaving intact parchment

(C) Shot holes in leaves

(D) **All of the above**

99. The cultivars showing resistance to DBM

(A) Pusa Drum Head (B) Golden Acre

(C) Pusa Mukta (D) **All of the above**

100. The intercrop recommended for DBM control in cabbage is

(A) **Tomato (1:1)** (B) Potato (1:1)

(C) Mustard (1:1) (D) Marigold (1:1)

101. The Trap crop recommended for DBM is

(A) Tomato (B) Potato

(C) **Mustard** (D) Marigold

102. The scientific name of cabbage head borer is

(A) *Pieris brasscicae* (B) *Plutella xylostella*

(C) *Crocidolomia binotalis* (D) ***Hellula undalis***

103. Thescientific name of cabbage leaf webber is

(A) *Pieris brasscicae* (B) *Plutella xylostella*

(C) ***Crocidolomia binotalis*** (D) *Hellula undalis*

104. The cabbage leaf webber lay eggs

(A) Upper side of leaves (B) Underside of leaves
(C) On shoot (D) In the soil

105. The cabbage leaf webber pupates in

(A) Upper side of leaves (B) Underside of leaves
(C) On shoot (D) **In the soil**

106. The over wintering stage of flea beetle is

(A) **Adult** (B) Larva
(C) Pupa (D) Egg

107. The ETL for Okra Jassid is

(A) **4-5 nymphs/plant** (B) 4-5 nymphs/leave
(C) 5-10 nymphs/leave (D) 5-10 nymphs/plant

108. The ETL for DBM is

(A) **10 larvae/ 20 plants** (B) 2-5 larvae/20 plants
(C) 10-20larvae/ 10 plant (D) 10 larvae / 10 plants

109. The Ber cultivars showing resistance to fruit fly is

(A) Sanaur-1 (B) Thornless
(C) Gola (D) **All of the above**

110. The Citrus cultivars showing resistance to Citrus leaf miner is

(A) Citrumelo (B) Morton
(C) Savage (D) **All of the above**

111. The Pomegranate cultivars showing resistance to fruit borer is

(A)Muscat (B) Speen
(C) G-137 (D) **All of the above**

112. The ETL for *Cosmopolites sordidus* in banana is

(A) **2 weevils/ Trap** (B) 4 weevils/trap
(C) 10 weevils/trap (D) 15 weevils / trap

113. The ETL for *Diaphorina citri* in citrus is

(A) **5-10 nymphs/leaf** (B) 10-15 nymphs/leaf
(C) 15-20 nymphs/leaf (D) 20 nymphs/leaf

114. The scientific name of mango hopper is

(A) *Idioscopus niveoparsus* (B) *Idioscopus clypealis*
(C) *Amritodus atkinsoni* (D) **All of the above**

115. The mango hopper is a

(A) **Monophagous pest**
(B) Oligophagous pest
(C) Polyphagous pest
(D) All of the above

116. The adults of mango hopper, *Idioscopus niveoparsus* can be identified by

(A) **Dark with wavy lines on wings and three spots on scutellum**
(B) Light brown with dark spots on the vertex and two spots on scutellum
(C) arge, light brown with two spots on scutellum
(D) None of the above

117. The adults of mango hopper, *Idioscopus clypealis* can be identified by

(A) Dark with wavy lines on wings and three spots on scutellum
(B) **light brown with dark spots on the vertex and two spots on scutellum**
(C) Large, light brown with two spots on scutellum
(D) None of the above

118. The adults of mango hopper, *Amritodus atkinsoni* can be identified by

(A) Dark with wavy lines on wings and three spots on scutellum
(B) Light brown with dark spots on the vertex and two spots on scutellum
(C) **Large, light brown with two spots on scutellum**
(D) None of the above

119. The Mango hopper lay eggs

(A) On the leaves
(B) **Inside the leaf tissue**
(C) On the fruits
(D) Inside the plants parts

120. The Honey dew excretion of Mango hopper results in the growth of fungi namely

(A) *Meliola mangiferae*
(B) *Capnodium mangiferum*
(C) **Both A and B**
(D) None of the above

121. The Mango stem borer belong to family

(A) **Cerambycidae**
(B) Bostrychidae
(C) Anthribidae
(D) None of the above

122. The Mango stem borer, *Batocera rufomaculata* lay eggs

(A) **Singly**
(B) In groups
(C) Egg mass
(D) None of the above

123. The Life cycle of mango stem borer, *Batocera rufomaculata* completes in

(A) 6 month
(B) 6-12 months
(C) 12-18 months
(D) **1-2 years**

124. The injection of 4 ml of methyl parathion 50 EC mixed in one litre water in the bore holes is done to control

(A) *Sternochaetus mangiferae* (B) ***Batocera rufomaculata***

(C) *Chlumetia transversa* (D) None of the above

125. The scientific name of mango nut weevil

(A) ***Sternochaetus mangiferae*** (B) *Batocera rufomaculata*

(C) *Chlumetia transversa* (D) None of the above

126. The scientific name of mango shoot borer

(A) *Sternochaetus mangiferae* (B) *Batocera rufomaculata*

(C) ***Chlumetia transversa*** (D) None of the above

127. The mango shoot borer, *Chlumetia transversa* belong to family

(A) Cerambycidae (B) Bostrychidae

(C) Anthribidae (D) **Noctuidae**

128. The scientific name of mango fruit fly is

(A) *Carpomyia vesuviana* (B) ***Bactocera dorsalis***

(C) *Bactocera cucrbitae* (D) None of the above

129. The parasitoids associated with mango fruit fly is

(A) *Opius compensatus* (B) *Dirhinus giffardi*

(C) *Spalangia phillippinensis* (D) **All of the above**

130. The mango mealy bug lay eggs

(A) On stem (B) On leaves

(C) **In the soil** (D) On all above plant parts

131. The Tree banding is done to control

(A) **Mango mealy bug** (B) Mango scale

(C) Mango hopper (D) All of the above

132. The scientific name of Mango mite is

(A) *Tetranychus urticae* (B) *Tetranychus cinnabarinus*

(C) *Tetranychus neocaledonics* (D) ***Aceria mangiferae***

133. The scientific name of Mango Gall midge is

(A) *Procystiphora mangiferae* (B) *Erosomyia indica*

(C) *Dasineura amaramanjarae* (D) **All of the above**

134. The dispause stage of Mango Gall midge is

(A) Larval in gall (B) **Pupal in soil**

(C) Adult on leaves (D) Egg on leaves

135. The dispause stage of Mango Gall midge is
   (A) Larval in gall (B) **Pupal in soil**
   (C) Adult on leaves (D) Egg on leaves

136. The Fruit sucking moth is a pest of
   (A) Apple (B) **Citrus**
   (C) Pomegranate (D) Ber

137. The damaging stage of fruit sucking moth is
   (A) **Adult** (B) Larva
   (C) Both Adult and larva (D) None of the above

138. The typical "Pin hole damage" in infested citrus fruits is the characteristic of
   (A) Citrus butterfly (B) **Fruit sucking moth**
   (C) Citrus psylla (D) None of the above

139. The larvae of fruit sucking moth is a
   (A) Caterpillar (B) Looper
   (C) **Semilooper** (D) Grub

140. The characteristic of larvae of fruit sucking moth is
   (A) Semilooper having a dorsal lump on the last segment of the body
   (B) Have distinct eye spots on head yellow / red lateral spots
   (C) When disturbed assumes a characteristic snake like posture
   (D) **All of the above**

141. The adults of fruit sucking moths can be managed by
   (A) Destruction of weeds/ alternate host plants (B) Bagging of fruits
   (C) bait killing (D) **All of the above**

142. The adults of fruit sucking moth, *Ophederes materna* is differentiated from other species by
   (A) **Three black spots on the fore wings**
   (B) Have white bands in the middle fore wing
   (C) Tripod black mark in the forewing and curved marking in hind wing
   (D) None of the above

143. The adults of fruit sucking moth, *Ophederes arcilla* is differentiated from other species by
   (A) Three black spots on the fore wings
   (B) **Have white bands in the middle fore wing**
   (C) Tripod black mark in the forewing and curved marking in hind wing
   (D) None of the above

144. The adults of fruit sucking moth, *Ophederes fullonica* is differentiated from other species by

(A) Three black spots on the fore wings

(B) Have white bands in the middle fore wing

(C) **Tripod black mark in the forewing and curved marking in hind wing**

(D) None of the above

145. The citrus leaf miner belong to family

(A) **Gracillaridae** (B) Noctuidae

(C) Pyraustidae (D) Pyralidae

146. The citrus leaf miner lay eggs

(A) Singly inside leaf tissue (B) **Singly on the leaf**

(C) Singly on the fruits (D) All of the above

147. The infestation of citrus leaf miner promotes which disease infection

(A) **Citrus canker** (B) Citrus scab

(C) LeafBlight (D) All of the above

148. The characteristic of larva of lemon butterfly is

(A) It resembles Bird dropping

(B) Has two reddish sac posterior to head for defence

(C) **Both A and B**

(D) None of the above

149. The citrus caterpillar lay eggs

(A) Singly (B) Groups of 2-5

(C) Egg mass (D) **Both A and B**

150. The egg parasitoid associated with Citrus caterpillar, *Papilio demoleus*

(A) *Trichogramma chilonis* (B) *Trichogramma brasiliensis*

(C) *Trichogramma acheae* (D) ***Trichogramma evanescens***

151. The scientific name of citrus psylla is

(A) ***Diaphirina citri*** (B) *Toxoptera citricidi*

(C) *Phyllocnistis citrella* (D) *Aleurocanthus woglumi*

152. The Vector of citrus greening disease is

(A) ***Diaphirina citri*** (B) *Toxoptera citricidi*

(C) *Phyllocnistis citrella* (D) *Aleurocanthus woglumi*

153. The scientific name of citrus Red scale is

(A) ***Aonidiella aurantii*** (B) *Toxoptera citricidi*

(C) *Phyllocnistis citrella* (D) *Aleurocanthus woglumi*

154. The scientific name of citrus cottony cushion scale is

(A) *Aonidiella aurantii* (B) ***Icerya purchsai***

(C) *Psudococcus filamentosus* (D) *Aleurocanthus woglumi*

155. The scientific name of citrus black fly is

(A) *Aonidiella aurantii* (B) *Icerya purchsai*

(C) *Psudococcus filamentosus* (D) ***Aleurocanthus woglumi***

156. The scientific name of citrus mealy bug is

(A) *Aonidiella aurantii* (B) *Icerya purchsai*

(C) ***Psudococcus filamentosus*** (D) *Aleurocanthus woglumi*

157. The scientific name of citrus leaf mite is

(A) ***Eutetranychus barksi*** (B) *Phyllocoptruta oleivora*

(C) *Oligonychus citri* (D) *Aleurocanthus woglumi*

158. The scientific name of citrus Fruit rust mite is

(A) *Eutetranychus barksi* (B) ***Phyllocoptruta oleivora***

(C) *Oligonychus citri* (D) *Aleurocanthus woglumi*

159. The scientific name of citrus nematode is

(A) *Meloiodogyne incognita* (B) *Ditylenchus angustatus*

(C) *Hoplolaimus indicus* (D) ***Tylenchulus semipenetrans***

160. The scientific name of citrus Lance nematode is

(A) *Meloiodogyne incognita* (B) *Ditylenchus angustatus*

(C) ***Hoplolaimus indicus*** (D) *Tylenchulus semipenetrans*

161. The major pest of guava is

(A) ***Bactrocera dorsalis*** (B) *Indarbela quadrinotata*

(C) *Saissetia coffea* (D) *Chloropulvinaria psidii*

162. The bark eating caterpillar of guava is

(A) *Bactrocera dorsalis* (B) *Indarbela quadrinotata*

(C) *Saissetia coffea* (D) *Chloropulvinaria psidii*

163. The composition of poison bait for fruit fly is

(A) **Malathion (20ml) + Molasses (500g) + water (2L)**

(B) Endosulfan (20ml) + Molasses (500g) + water (5L)

(C) Carbaryl (20ml) + Molasses (500g) + water (2L)
(D) None of the above

164. The *Deudorix isocrates* belong to family is
(A) Papilionidae (B) Metarbelidae
(C) **Lycaenidae** (D) Pyralidae

165. The *Deudorix isocrates* is a major pest of
(A) **Pomegranate** (B) Ber
(C) Citrus (D) Loquat

166. The *Deudorix isocrates* lay eggs
(A) Singly on the calyx (B) Singly on the small fruits
(C) Singly on the leaves (D) **Both A and B**

167. The Lac insect, *Kerria lacca* is a pest of
(A) **Ber** (B) Guava
(C) Loquat (D) Banana

168. The predator of Lac insect is
(A) *Eublemma amabilis* (B) *Holocera pulverea*
(C) **Both A and B** (D) None of the above

169. The fruit fly infesting Ber is
(A) *Bactrocera dorsalis* (B) *Bactrocera zonatus*
(C) ***Carpomyia vesuviana*** (D) *Bactrocera tau*

170. The fruit fly infesting Ber is active during the season
(A) Summer (B) Autumn
(C) Rainy season (D) **Winter**

171. The *Sthenias grisator* belong to family
(A) **Cerambycidae** (B) Noctuidae
(C) Bostrychidae (D) Metarbelidae

172. The Ringing or Girdling of grapevine is due to the attack of
(A) ***Sthenias grisator*** (B) *Scelodonta strigicollis*
(C) *Sylepta lunalis* (D) *Sinoxylon anale*

173. The Shot hole symptom in grapevine is due to the attack of
(A) *Sthenias grisator* (B) ***Scelodonta strigicollis***
(C) *Sylepta lunalis* (D) *Sinoxylon anale*

174. The scientific name of grapevine leaf roller is due to the attack of
   (A) *Sthenias grisator* (B) *Scelodonta strigicollis*
   (C) ***Sylepta lunalis*** (D) *Sinoxylon anale*

175. The grapevine leaf roller belong to family
   (A) Cerambycidae (B) Noctuidae
   (C) Bostrychidae (D) **Pyralidae**

176. The grapevine beetle, *Sinoxylon anale* belong to family
   (A) Cerambycidae (B) Noctuidae
   (C) **Bostrychidae** (D) Metarbelidae

177. The parasitoid for controlling papaya mealy bug is
   (A) *Pseudolaptomastyx mexicana* (B) *Acerophagus papaye*
   (C) *Anagyrus loecki* (D) **All of the above**

178. The scientific name of Banana rhizome weevil is
   (A) *Odioporus longicollis* (B) ***Cosmpolites sordidus***
   (C) *Nacoleia octasema* (D) *Pentalonia nigeronervosa*

179. The scientific name of Banana Pseudostem borer is
   (A) ***Odioporus longicollis*** (B) *Cosmpolites sordidus*
   (C) *Nacoleia octasema* (D) *Pentalonia nigeronervosa*

180. The scientific name of Banana Aphid is
   (A) *Odioporus longicollis* (B) *Cosmpolites sordidus*
   (C) *Nacoleia octasema* (D) ***Pentalonia nigeronervosa***

181. The scientific name of Banana scale moth is
   (A) *Odioporus longicollis* (B) *Cosmpolites sordidus*
   (C) ***Nacoleia octasema*** (D) *Pentalonia nigeronervosa*

182. The scientific name of Banana rhizome weevil is
   (A) *Odioporus longicollis* (B) ***Cosmpolites sordidus***
   (C) *Nacoleia octasema* (D) *Pentalonia nigeronervosa*

183. The Black head disease/Toppling disease in banana is caused by
   (A) Insect-pest (B) Virus
   (C) **Nematode** (D) Mite

184. The Black head disease/Toppling disease in banana is caused by
   (A) *Odioporus longicollis* (B) ***Rahdopholus similis***
   (C) *Nacoleia octasema* (D) *Pentalonia nigeronervosa*

185. The vector of bunchy top disease in banana is
   (A) *Odioporus longicollis* (B) *Cosmpolites sordidus*
   (C) *Nacoleia octasema* (D) ***Pentalonia nigeronervosa***

186. The scientific name of Banana rhizome weevil is
   (A) *Odioporus longicollis* (B) ***Cosmpolites sordidus***
   (C) *Nacoleia octasema* (D) *Pentalonia nigeronervosa*

187. The damaging stage of Banana tingid is
   (A) Nymphs (B) Adults
   (C) **Both nymph and adult** (D) None of the above

188. The scientific name of Fig Midge is
   (A) *Odioporus longicollis* (B) *Rhyncaphytopus ficifoliae*
   (C) ***Anjeerodiplosis peshawarensis*** (D) *Pentalonia nigeronervosa*

189. The scientific name of Jackfruit leaf webber is
   (A) ***Perina nuda*** (B) *Rhyncaphytopus ficifoliae*
   (C) *Anjeerodiplosis peshawarensis* (D) *Acrocercops phaeospora*

190. The scientific name of Jamun leaf miner is
   (A) *Perina nuda* (B) *Rhyncaphytopus ficifoliae*
   (C) *Anjeerodiplosis peshawarensis* (D) ***Acrocercops phaeospora***

191. The scientific name of Tamarind fruit borer is
   (A) *Perina nuda* (B) ***Phycita orthoclina***
   (C) *Anjeerodiplosis peshawarensis* (D) *Acrocercops phaeospora*

192. The scientific name of Litchi bug is
   (A) *Perina nuda* (B) *Rhyncaphytopus ficifoliae*
   (C) *Anjeerodiplosis peshawarensis* (D) ***Chrysocoris stolii***

193. The scientific name of Litchi mite is
   (A) *Tetranychus neocaledonics* (B) *Aceria lycopersici*
   (C) ***Aceria litchii*** (D) *Tetranychus cinnabarinus*

194. The scientific name of Sapota leaf webber is
   (A) *Perina nuda* (B) *Rhyncaphytopus ficifoliae*
   (C) ***Nephopteryx eugraphella*** (D) *Chrysocoris stolii*

195. The ETL for grapevine flea beetle is
   (A) **2% foliar damage** (B) 4% foliar damage
   (C) 6% foliar damage (D) 10% foliar damage

196. The ETL for grapevine Mealy bug is

(A) **1% bunch infestation** (B) 5% bunch infestation
(C) 10% bunch infestation (D) 15% foliar damage

197. The ETL for mango leaf webber is

(A) **5% infestation** (B) 10% infestation
(C) 15% infestation (D) 20% infestation

198. The ETL for Mango hopper is

(A) **5 adults/panicle** (B) 5-10 adults/panicle
(C) 10-15 adults/panicle (D) 15-20 adults/panicle

199. The ETL for Ber fruit fly is

(A) **1-2% incidence** (B) 2-5% incidence
(C) 4-8% incidence (D) 5-10% incidence

200. For controlling Citrus mealybug, *Cryptolaemus montrouzieri* is released at the rate of

(A) 5 beetles/Plant (B) **10 beetles/Plant**
(C) 15 beetles/Plant (D) None of the above

201. For controlling Grape mealy bug, *Cryptolaemus montrouzieri* is released at the rate of

(A) 5 beetles/vine (B) **10 beetles/vine**
(C) 15 beetles/vine (D) None of the above

202. The plum hairy caterpillar, *Euproctis fraterna* is a major pest of

(A) Ber (B) **Phalsa**
(C) Guava (D) None of the above

203. The damaging stage of citrus black fly is

(A) **Maggot** (B) Grub
(C) Wriggler (D) None of the above

204. The citrus blossom midge belong to family

(A) Margarodidae (B) **Ceccidomyidae**
(C) Coccidae (D) None of the above

205. The scaly appearance of apple fruits surrounded by pink, scarlet or red colour area indicate the attack of

(A) San Jose Scale (B) Cottony cushion scale
(C) Apple wooly aphid (D) Apple coddling moth

206. *Trichogramma embryophagum* is a parasitoid of

(A) San Jose Scale (B) Cottony cushion scale

(C) Apple wooly aphid (D) **Apple coddling moth**

207. The San Jose Scale overwinter in

(A) **Nymphal stage** (B) Egg stage

(C) Adult stage (D) All of the above

208. The parasitoid of San Jose Scale is

(A) *Aphytis* sp. (B) *Encarsia perniciosi*

(C) **Both A and B** (D) *Aphelinus mali*

209. To control San Jose Scale, *Encarsia perniciosi* is released at the rate of

(A) 500 adults/infested plant (B) 1000 adults/infested plant

(C) 1000 adults/infested plant (D) **2000 adults/infested plant**

210. The insect having quarantine significance is

(A) San Jose Scale (B) Cottony cushion scale

(C) Apple wooly aphid (D) **Apple coddling moth**

211. The ESSO tree spray oil emulsion is recommended for the control of

(A) **San Jose Scale** (B) Cottony cushion scale

(C) Apple wooly aphid (D) Apple coddling moth

212. The Rootstock resistance against Wooly apple aphid

(A) Morton Stock-778 (B) Northern Spy

(C) Golden Delicious (D) **All of the above**

213. The damage symptoms of Wooly apple aphid is

(A) Gall formation on roots (B) Whitish cottony patches on plant parts

(C) Yellowing of leaves (D) **All of the above**

214. The Scientific name of Wooly apple aphid is

(A) *Quadraspidiotus pernicious* (B) ***Eriosoma lanigerum***

(C) *Dorysthenes hugelii* (D) *Cydia pomonella*

215. The Winter and summer migration is characteristic of

(A) *Quadraspidiotus pernicious* (B) ***Eriosoma lanigerum***

(C) *Dorysthenes hugelii* (D) *Cydia pomonella*

216. The parasitoid of Wooly apple aphid is

(A) *Aphytis* sp. (B) *Encarsia perniciosi*

(C) Both A and B (D) ***Aphelinus mali***

217. The Codling moth lay eggs

(A) **Singly** (B) In groups
(C) Egg mass (D) None of the above

218. The First brood/generation of Codling moth enter the fruit through

(A) **Calyx** (B) The side
(C) Pedicel (D) All of the above

219. The Second brood/generation of Codling moth enter the fruit through

(A) Calyx (B) **Theside**
(C) Pedicel (D) All of the above

220. The Scientific name of Apple stem borer is

(A) *Quadraspidiotus pernicious* (B) ***Apriona cinerea***
(C) *Dorysthenes hugelii* (D) *Cydia pomonella*

221. The Scientific name of Apple root borer is

(A) *Quadraspidiotus pernicious* (B) *Eriosoma lanigerum*
(C) ***Dorysthenes hugelii*** (D) *Malacosoma indicum*

222. The Scientific name of Tent caterpillar is

(A) *Quadraspidiotus pernicious* (B) *Eriosoma lanigerum*
(C) *Dorysthenes hugelii* (D) ***Malacosoma indicum***

223. The Tent caterpillar belong to family

(A) Buperstidae (B) **Lasiocampidae**
(C) Pyraustidae (D) Crambidae

224. The damage symptoms Peach leaf curl aphid resembles with those caused by

(A) ***Taphrina deformans*** (B) *Alternaria* sp.
(C) *Pythium apahnidermatum* (D) None of the above

225. The Scientific name of Peach twig borer is

(A) ***Anarisa lineatella*** (B) *Sphenoptera laferti*
(C) *Aeolesthes holoserica* (D) *Alcidodes porrectirostris*

226. The Peach twig borer belong to family

(A) Cerambycidae (B) Buperstidae
(C) **Gelechiidae** (D) Chrysomelidae

227. The Peach Stem borer belong to family

(A) Cerambycidae (B) **Buperstidae**
(C) Gelechiidae (D) Chrysomelidae

228. The bored holes with gum globules oozing out from the stem is due to the attack of
    (A) *Anarisa lineatella* (B) ***Sphenoptera laferti***
    (C) *Dorysthenes hugelii* (D) *Malacosoma indicum*

229. The Scientific name of Walnut wevil is
    (A) *Anarisa lineatella* (B) *Sphenoptera laferti*
    (C) *Aeolesthes holoserica* (D) ***Alcidodes porrectirostris***

230. The Scientific name of Cherry stem borer is
    (A) *Anarisa lineatella* (B) *Sphenoptera laferti*
    (C) ***Aeolesthes holoserica*** (D) *Alcidodes porrectirostris*

231. The damaging stage of walnut weevil is
    (A) Grub (B) Adult
    (C) **Both Grub and Adult** (D) None of the above

232. The Overwintering stage of walnut weevil is
    (A) Grub (B) **Adult**
    (C) Both Grub and Adult (D) None of the above

233. The Scientific name of Gypsy moth is
    (A) ***Lymantria obfuscata*** (B) *Sphenoptera laferti*
    (C) *Aeolesthes holoserica* (D) *Alcidodes porrectirostris*

234. The Scientific name of Cherry stem borer is
    (A) *Anarisa lineatella* (B) ***Myllocerus lactivirens***
    (C) *Aeolesthes holoserica* (D) *Alcidodes porrectirostris*

235. The Scientific name of Areca nut mirid bug is
    (A) ***Carvalhoia arecae*** (B) *Plocaederus ferrugineus*
    (C) *Aprocercops syngramma* (D) *Helopeltis antonii*

236. The Scientific name of Cashew nut tree borer is
    (A) *Carvalhoia arecae* (B) ***Plocaederus ferrugineus***
    (C) *Aprocercops syngramma* (D) *Helopeltis antonii*

237. The Scientific name of Cashew leaf miner is
    (A) *Carvalhoiaarecae* (B) *Plocaederus ferrugineus*
    (C) ***Aprocercops syngramma*** (D) *Nephopteryx eugraphylla*

238. The Scientific name of Sapota leaf webber/ Chickoo Moth is
    (A) *Carvalhoia arecae* (B) *Plocaederus ferrugineus*
    (C) *Aprocercops syngramma* (D) ***Nephopteryx eugraphylla***

239. The Sapota leaf webber/ Chickoo Moth belong to family
(A) Cerambycidae (B) **Pyraustidae**
(C) Gelechiidae (D) Chrysomelidae

240. The Coconut Rhinoceros beetle belong to family
(A) Cerambycidae (B) Buperstidae
(C) **Scarabaeidae** (D) Chrysomelidae

241. The damaged leaves show characteristic 'V' shaped clippings or holes in the leaflets in coconut is due to the attack of
(A) **Coconut Rhinoceros beetle** (B) Red Palm weevil
(C) Coconut whitegrub (D) Black headed caterpillar

242. The gummosis, crown toppling and drying up of young heart leaves in coconut is due to the attack of
(A) Coconut Rhinoceros beetle (B) **Red Palm weevil**
(C) Coconut whitegrub (D) Black headed caterpillar

243. The most damaging pest in coconut is
(A) **Coconut Rhinoceros beetle** (B) Red Palm weevil
(C) Coconut whitegrub (D) Black headed caterpillar

244. The gummosis, crown toppling and drying up of young heart leaves in coconut is due to the attack of
(A) Coconut Rhinoceros beetle (B) **Red Palm weevil**
(C) Coconut whitegrub (D) Black headed caterpillar

245. The Red ring disease of coconut palm is caused by
(A) *Aphelenchoides ritzembosi* (B) ***Rhadinaphelenchus cocophilus***
(C) *Anguina tritici* (D) *Pratylenchus citrophilus*

246. The Red ring disease of coconut palm is caused by
(A) Mites (B) **Nematodes**
(C) Fungi (D) Insect pest

247. The Scientific name of Coconut Rhinoceros beetle is
(A) *Rhynchophorus ferrugineus* (B) *Opisina arenosella*
(C) *Leucopholis coneophora* (D) ***Oryctes rhinoceros***

248. The Scientific name of Coconut Red palm weevil is
(A) ***Rhynchophorus ferrugineus*** (B) *Opisina arenosella*
(C) *Leucopholis coneophora* (D) *Oryctes rhinoceros*

249. The Scientific name of Coconut white grub is

(A) *Rhynchophorus ferrugineus* (B) *Opisina arenosella*

(C) ***Leucopholis coneophora*** (D) *Oryctes rhinoceros*

250. The Scientific name of Coconut Black headed caterpillar is

(A) *Rhynchophorus ferrugineus* (B) ***Opisina arenosella***

(C) *Leucopholis coneophora* (D) *Oryctes rhinoceros*

251. The Scientific name of Coconut mite is

(A) *Rhynchophorus ferrugineus* (B) ***Aceria guerreronis***

(C) *Leucopholis coneophora* (D) *Oryctes rhinoceros*

252. The natural enemy of Coconut Rhinoceros beetle is

(A) ***Platymeris laevocollis*** (B) *Scotia erratica*

(C) *Goniozus nephantidis* (D) *Chilocoris nigrita*

253. The natural enemy of Coconut black headed caterpillar is

(A) *Platymeris laevocollis* (B) *Scotia erratica*

(C) ***Goniozus nephantidis*** (D) *Chilocoris nigrita*

254. To control Coconut Black headed caterpillar, *Goniozus nephantidis* is release at the rate of

(A) **10 adults/palm** (B) 20 adults/palm

(C) 30 adults/palm (D) 40 adults/palm

255. Mixing of *Metarhizium anisopliae* in manure pits is recommended for the control of

(A) *Rhynchophorus ferrugineus* (B) ***Oryctes rhinoceros***

(C) *Leucopholis coneophora* (D) *Oryctes rhinoceros*

256. Pheromone traps "Ferrolure" is recommended for the control of

(A) ***Rhynchophorus ferrugineus*** (B) *Oryctes rhinoceros*

(C) *Leucopholis coneophora* (D) *Oryctes rhinoceros*

257. The scientific name of coffee stem borer is

(A) *Coccus viridis* (B) ***Xylotrechus quadripes***

(C) *Hypothenemus hampei* (D) *Xylosandrus compactus*

258. The scientific name of coffee Berry borer is

(A) *Coccus viridis* (B) *Xylotrechus quadripes*

(C) ***Hypothenemus hampei*** (D) *Xylosandrus compactus*

259. The scientific name of coffee Shot Hole borer is

(A) *Coccus viridis* (B) *Xylotrechus quadripes*

(C) *Hypothenemus hampei* (D) ***Xylosandrus compactus***

260. The scientific name of coffee green scale is

(A) ***Coccus viridis*** (B) *Xylotrechus quadripes*

(C) *Hypothenemus hampei* (D) *Xylosandrus compactus*

261. The coffee stem borer belong to family

(A) Coccidae (B) Scolytidae

(C) Chrysomelidae (D) **Cerambycidae**

262. The coffee Shot Hole borer belong to family

(A) Coccidae (B) **Scolytidae**

(C) Chrysomelidae (D) Cerambycidae

263. The coffee Berry borer belong to family

(A) Coccidae (B) **Scolytidae**

(C) Chrysomelidae (D) Cerambycidae

264. The scientific name of coffee Striped tail mealy bug is

(A) *Coccus viridis* (B) ***Ferrisia virigata***

(C) *Hypothenemus hampei* (D) *Xylosandrus compactus*

265. The Tea mosquito bug belong to family

(A) Culicidae (B) Coccidae

(C) Tingidae (D) **Miridae**

266. The scientific name of Tea mosquito bug is

(A) ***Helopeltis theivora*** (B) *Andraca bipunctata*

(C) *Zeuzera coffeae* (D) *patulicraspeda castaneiceps*

267. The scientific name of Tea bunch caterpillar is

(A) *Helopeltis theivora* (B) ***Andraca bipunctata***

(C) *Zeuzera coffeae* (D) *Spatulicraspeda castaneiceps*

268. The scientific name of Tea Humped slug caterpillar is

(A) *Helopeltis theivora* (B) *Andraca bipunctata*

(C) *Zeuzera coffeae* (D) ***Spatulicraspeda castaneiceps***

269. The scientific name of Tea Red borer bug is

(A) *Helopeltis theivora* (B) *Andraca bipunctata*

(C) ***Zeuzera coffeae*** (D) *Spatulicraspeda castaneiceps*

270. The Tea Red borer belong to family
(A) **Cossidae** (B) Bombycidae
(C) Cochlididae (D) Miridae

271. The Tea Humped slug caterpillar belong to family
(A) Cossidae (B) Bombycidae
(C) **Cochlididae** (D) Miridae

272. The Tea Bunch caterpillar belong to family
(A) Cossidae (B) **Bombycidae**
(C) Cochlididae (D) Miridae

273. The scientific name of Tea Red spider mite is
(A) *Poyphagotarsonemus latus* (B) ***Oligonychus coffeae***
(C) *Brevipalpus phoenicis* (D) *Calaerus theae*

274. The scientific name of Red Crevice Tea mite is
(A) *Poyphagotarsonemus latus* (B) *Oligonychus coffeae*
(C) ***Brevipalpus phoenicis*** (D) *Calaerus theae*

275. The scientific name of Yellow Tea mite is
(A) ***Poyphagotarsonemus latus*** (B) *Oligonychus coffeae*
(C) *Brevipalpus phoenicis* (D) *Calaerus theae*

276. The scientific name of Purple Tea mite is
(A) *Poyphagotarsonemus latus* (B) *Oligonychus coffeae*
(C) *Brevipalpus phoenicis* (D) ***Calaerus theae***

277. The scientific name of Pink Tea mite is
(A) *Poyphagotarsonemus latus* (B) ***Acaphylla carinus***
(C) *Brevipalpus phoenicis* (D) *Calaerus theae*

278. The Red spider mite of Tea belong to family
(A) Tarsonemidae (B) **Tetranychidae**
(C) Tenuipalpidae (D) Eriophyidae

279. The Pink Mite of Tea belong to family
(A) Tarsonemidae (B) Tetranychidae
(C) Tenuipalpidae (D) **Eriophyidae**

280. The Purple mite of Tea belong to family
(A) Tarsonemidae (B) Tetranychidae
(C) Tenuipalpidae (D) **Eriophyidae**

281. The Yellow Tea mite belong to family
(A) **Tarsonemidae** (B) Tetranychidae
(C) Tenuipalpidae (D) Eriophyidae

282. The Red Crevice tea mite belong to family
(A) Tarsonemidae (B) Tetranychidae
(C) **Tenuipalpidae** (D) Eriophyidae

283. The *Aestherastis circulata* is a major pest of
(A) Coffee (B) **Rubber**
(C) Red palm (D) Teak

284. The Insect vector of cardamom disease/Katte disease is
(A) Aspidiotus destructor (B) ***Pentalonia nigreonervosa***
(C) *Aphis craccivora* (D) *Scirtothrips cardamomi*

285. The Insect vector of *Amomum*Mosiac or Foorky disease of large cardamom is
(A) *Aspidiotus destructor* (B) ***Pentalonia nigreonervosa***
(C) *Aphis craccivora* (D) *Scirtothrips cardamomi*

286. The castor capsule borer, *Conogethes punctiferalis* is a major pest of
(A) Coffee (B) Rubber
(C) Palm (D) **Cardamom**

287. The hairy caterpillar infesting cardamom is
(A) *Eupterote canarica* (B) *Lenodera vittata*
(C) *Eupterote cardamomi* (D) **All of the above**

288. The rhizome weevil infesting cardamom is
(A) *Eupterote canarica* (B) *Lenodera vittata*
(C) *Eupterote cardamomi* (D) ***Prodioctes haemiaticus***

289. The Insect vector of *Chirke* disease of cardamom is
(A) *Aspidiotus destructor* (B) *Pentalonia nigreonervosa*
(C) *Aphis craccivora* (D) ***Rhophalosiphum maidis***

290. The Pollu beetle is a specific pest of
(A) Cinnamon (B) **Black pepper**
(C) Turmeric (D) Ginger

291. The damaging stage of Pollu beetle is
(A) Adults (B) Grub
(C) **Both Adult and grub** (D) None of the above

292. The Pollu beetle belong to family

(A) **Chrysomelidae** (B) Meloidae

(C) Buperstidae (D) Cerambycidae

293. The Pollu beetle lay eggs

(A) On the berries (B) **In the berries**

(C) On the leaf (D) In the soil

294. The Pollu beetle pupates

(A) On the berries (B) In the berries

(C) On the leaf (D) **In the soil**

295. The skipper butterfly, *Udaspes folus* is a major pest of

(A) Cinnamon (B) Blackpepper

(C) **Turmeric** (D) Ginger

296. The skipper butterfly, *Udaspes folus* belong to family

(A) Pieridae (B) **Hesperiidae**

(C) Pyralidae (D) Noctuidae

297. The scale,*Aspidiotus hartii* is a major pest of

(A) Cinnamon (B) Blackpepper

(C) Turmeric (D) **Ginger**

298. The scale,*Aspidiotus hartii* belong to family

(A) Coccidae (B) Pseudococcidae

(C) **Diaspididae** (D) Miridae

299. The *Bemesia tabaci* is a major pest of

(A) Cinnamon (B) Blackpepper

(C) **Coriander** (D) Ginger

300. The butterfly,*Chilasia clytia* is a major pest of

(A) **Cinnamon** (B) Blackpepper

(C) Turmeric (D) Ginger

301. The Cinnamon butterfly,*Chilasia clytia* belong to family

(A) Pieridae (B) Hesperiidae

(C) Pyralidae (D) **Papilionidae**

302. To control plant parasitic nematodes, *Paeciliomyces lilacinus* is recommended at the rate of

(A) 10 kg/ha (B) **15kg/ha**

(C) 20 kg/ha (D) 25 kg/ha

303. To control *Oryctes rhinoceros*, Baculovirus is recommended at the rate of
(A) **10 infected beetle/palm** (B) 15 infected beetle/palm
(C) 20 infected beetle/palm (D) 25 infected beetle/palm

304. The *Macrosiphum miscanthi* and *M. rosaeformis* are pests of
(A) **Rose** (B) Jasmine
(C) Sunflower (D) Chrysanthemum

305. The *Macrosiphum miscanthi* and *M. rosaeformis* belong to family
(A) Coccidae (B) **Aphididae**
(C) Jassidae (D) Psyllidae

306. The Tingid bug, *Urentis euonymus* is a pests of
(A) Rose (B) Jasmine
(C) **Hollyhock** (D) Chrysanthemum

307. The Tingid bug, *Urentis euonymus* overwinter in
(A) **Egg stage** (B) Nymphal stage
(C) Adult stage (D) All of the above

308. The Sunflower lacewing bug scientific name is
(A) *Urentis euonymus* (B) *Macrosiphum miscanthi*
(C) ***Cadmilos retarius*** (D) *Danais chrysippus*

309. The Ak butterfly/African Monarch/Plain tiger scientific name is
(A) *Urentis euonymus* (B) *Macrosiphum miscanthi*
(C) *Cadmilos retarius* (D) ***Danais chrysippus***

310. The thrips infesting Jasmine is
(A) *Rhipihorothrips cruentatus* (B) ***Thrips orientalis***
(C) *Thrips tabaci* (D) *Scirtothrips dorsalis*

311. The Leaf web worm *Nausinoe geometralis* is a pest of
(A) Rose (B) **Jasmine**
(C) Hollyhock (D) Chrysanthemum

312. The Leaf web worm *Nausinoe geometralis* belong to the family
(A) Crambidae (B) **Pyraustidae**
(C) Pyralidae (D) Noctuidae

313. The Jasmine Bud worm, *Hendecasis duiplifascilis* belong to the family
(A) Crambidae (B) **Pyraustidae**
(C) Pyralidae (D) Noctuidae

314. The Jasmine Gallery worm, *Elasmopalpus jasminophagous* belong to the family

(A) Crambidae (B) Pyraustidae

(C) **Pyralidae** (D) Noctuidae

315. The sporadic but specific pest of Lily is

(A) *Nausinoe geometralis* (B) ***Polytela gloriosa***

(C) *Hendecasis duiplifascilis* (D) *Danais chrysippus*

316. The Lily moth,*Polytela gloriosa* belong to the family

(A) Crambidae (B) Pyraustidae

(C) Pyralidae (D) **Noctuidae**

317. The major pest of Gerbera is

(A) ***Liriomyza trifolii*** (B) *Polytela gloriosa*

(C) *Hendecasis duiplifascilis* (D) *Danais chrysippus*

318. The mites infesting Jasmine is

(A) ***Aceria jasmini*** (B) *Tetranychus urticae*

(C) *Aceria cajani* (D) *Polyphagotarsonemus latus*

Unit-11

Storage Entomology

MCQ's

1. The book "*Stored Grain Pests and Their Management*" is authored by .
   (A) **BP Khare** (B) DW Hall
   (C) Subramanyam B & Hagstrum DW (D) V Khader

2. The author of "*Handling and Storage of Food Grains in Tropical and Subtropical Areas*" is
   (A) BP Khare (B) **DW Hall**
   (C) Subramanyam B & Hagstrum DW (D) V Khader

3. The author of "*Interrelated Management of Insects in Stored Products*" is
   (A) BP Khare (B) DW Hall
   (C) **Subramanyam B & Hagstrum DW** (D) V Khader

4. The estimated losses in stored food grains as per FAO is
   (A) 5-8% (B) **8-10%**
   (C) 10-15% (D) 15-20%

5. The optimum temperature for growth and development of stored pests is
   (A) 18-28°C (B) 22-28°C
   (C) **28-32°C** (D) 30-35°C

6. The insect pests associated with hidden infestation is
   (A) Paddy moth (B) Pulse beetle
   (C) **Both A and B** (D) None of the above

7. The infestation of fresh stock from the pest already presented in the godown is
   (A) Hidden infestation (B) **Cross infestation**
   (C) Both A and B (D) None of the above

8. The physical method for detection of infestation of stored pest
   (A) NMR and IR spectroscopy (B) Berleese method
   (C) Floatation method (D) **All of the above**
9. The Chemical method for detection of infestation of stored pest
   (A) ELISA test (B) Uric acid anaysis
   (C) Specific gravity method (D) **All of the above**
10. The staining technique for detection of infestation of stored pest
   (A) Ninhydrin method (B) Egg plugs
   (C) **Both A and B** (D) None of the above
11. The physical method for detection of infestation of stored pest
   (A) NMR and IR spectroscopy (B) Berleese method
   (C) Floatation method (D) **All of the above**
12. The presence of exit holes in whole spices indicate the infestation of
   (A) *Ephestia cautella* (B) ***Stegobium paniceum***
   (C) *Sitophilus oryzae* (D) *Callosobruchus*sp.
13. The presence of pupal cases sticking to peanut shells and gunny bags indicate the infestation of
   (A) *Ephestia cautella* (B) *Stegobium paniceum*
   (C) *Sitophilus oryzae* (D) ***Caryedon serratus***
14. The stored pests which attack sound grains are known as
   (A) Secondary pest (B) **Primary pest**
   (C) Tertiary pest (D) None of the above
15. The stored pests which attack broken grains are known as
   (A) **Secondary pest** (B) Primary pest
   (C) Detritus pest (D) None of the above
16. The stored pests which attack unhealthy grains are known as
   (A) Secondary pest (B) Primary pest
   (C) **Detritus pest** (D) None of the above
17. Which of the following is a detritus stored pest
   (A) *Ephestia cautella* (B) *Sitotroga cerelella*
   (C) *Sitophilus oryzae* (D) **Corn sap beetle**
18. Which of the following is a primary stored pest
   (A) *Ephestia cautella* (B) *Sitotroga cerelella*
   (C) *Sitophilus oryzae* (D) **All of the above**

19. Which of the following is a primary stored pest

(A) *Rhizopertha dominica* (B) *Callosobruchus* sp.

(C) *Corcyra cephalonica* (D) **All of the above**

20. Which of the following is a secondary stored pest

(A) *Tribolium castaneum* (B) *Cryptolestes ferrugineus*

(C) *Oryzaephilus surinamensis* (D) **All of the above**

21. The rice weevil can be identified by

(A) Having snout/beak (B) Antennae Geniculate

(C) Four reddish or yellowish spot on elytra (D)**All of the above**

22. The rice weevil can be identified by

(A) Having snout/beak (B) Antennae Geniculate

(C) Four reddish or yellowish spot on elytra (D) **All of the above**

23. The typical damaging symptom of rice weevil is

(A) **Circular exit hole** (B) Irregular exit hole

(C) Split grains (D) None of the above

24. The typical damaging symptom of Lesser grain borer is

(A) Circular exit hole (B) **Irregular exit hole**

(C) Split grains (D) None of the above

25. The damaging stage of rice weevil is

(A) Adultonly (B) Larva only

(C) **Both Larva and adults** (D) None of the above

26. The Lesser grain borer can be identified by

(A) Head deflexed downward

(B) prominent constriction between prothorax and elytra

(C) antenna 10 segmented ending in tripartite club

(D) **All of the above**

27. The damaging stage of Lesser grain borer is

(A) Adultonly (B) Larva only

(C) **Both Larva and adults** (D) None of the above

28. The Family of Lesser grain borer is

(A) **Bostrichidae** (B) Tenebrionidae

(C) Dermestidae (D) Bruchidae

29. The Family of Khapra beetle is
   (A) Bostrichidae (B) Tenebrionidae
   (C) **Dermestidae** (D) Bruchidae
30. The Family of Red flour beetle is
   (A) Bostrichidae (B) **Tenebrionidae**
   (C) Dermestidae (D) Bruchidae
31. The scientific name of Khapra beetle is
   (A) *Rhizopertha dominica* (B) ***Trogoderma granarium***
   (C) *Tribolium csataneum* (D) *Sitophilus oryzae*
32. The scientific name of Lesser grain borer is
   (A) ***Rhizopertha dominica*** (B) *Trogoderma granarium*
   (C) *Tribolium csataneum* (D) *Sitophilus oryzae*
33. The damaging stage of Khapra beetle is
   (A) Adultonly (B) **Larvaonly**
   (C) Both Larva and adults (D) None of the above
34. The hibernating stage of Khapra beetle is
   (A) Adultonly (B) **Larvaonly**
   (C) Both Larva and adults (D) None of the above
35. The Larvae of Khapra beetle is
   (A) **Positively thigmotactic** (B) Negatively thigmotactic
   (C) Thigmotactic neutral (D) None of the above
36. Which of these is a strong flier causing cross infestation
   (A) **Lesser grain borer** (B) Khapra beetle
   (C) Rice weevil (D) Rice weevil
37. The stored pest capable of surviving under low moisture (<4%) of seed
   (A) Lesser grain borer (B) **Khapra beetle**
   (C) Rice weevil (D) Rice weevil
38. Which of these is a strong flier causing cross infestation
   (A) **Lesser grain borer** (B) Khapra beetle
   (C) Red rust flour beetle (D) Rice weevil
39. The Khapra beetle can be identified by
   (A) Adults having no distinction of head, thorax and abdomen
   (B) Antennae clubbed
   (C) Larva have tuft forming tail at posterior end
   (D) **All of the above**

40. The Oxygen requirement for survival of Khapra beetle eggs is

(A) 19.8% (B) 18.8%

(C) **16.8%** (D) 14.8%

41. Which of these shows cannibalism

(A) Lesser grain borer (B) Khapra beetle

(C) **Red rust flour beetle** (D) Rice weevil

42. Which of Tribolium species is a flyer

(A) ***Tribolium castaneum*** (B) *Tribolium confusum*

(C) Both A and B (D) None of the above

43. The damaging stage of Red rust flour beetle is

(A) Adult only (B) Larva only

(C) **Both Larva and adults** (D) None of the above

44. The offensive pungent smell of infested grain due to Red rust flour beetle is because of

(A) **Gaseous quinones** (B) Gaseous phenols

(C) Gaseous acids (D) None of the above

45. The Red rust flour beetle can be identified by

(A) All three body segments are distinctly visible dorsally

(B) Antennae clubbed

(C) Larva negatively phototactic

(D) **All of the above**

46. The hibernating stage of Pulse beetle is

(A) Adultonly (B) **Larvaonly**

(C) Both Larva and adults (D) None of the above

47. Which of the following is both field and storage pest

(A) *Chromatomyia horticola* (B) *Lampides boeticus*

(C) *Ophiomyia phaseoli* (D) ***Callosobruchus chinensis***

48. The Pulse beetle can be identified by

(A) Spine on each of the inner and outer edges of the of hind femur

(B) Antennae Serrate

(C) Elytra truncate not covering the pygidium

(D) **All of the above**

49. The scientific name of the spotted bean bruchid is
    (A) *Callosobruchus chinensis* (B) ***Callosobruchus maculatus***
    (C) *Callosobruchus phaseoli* (D) *Callosobruchus analis*

50. The scientific name of the Lentil bruchid/Dhora is
    (A) *Callosobruchus chinensis* (B) *Callosobruchus maculatus*
    (C) ***Callosobruchus phaseoli*** (D) *Callosobruchus analis*

51. The scientific name of the Moong bruchid/dhora is
    (A) *Callosobruchus chinensis* (B) *Callosobruchus maculatus*
    (C) *Callosobruchus phaseoli* (D) ***Callosobruchus analis***

52. The scientific name of the Pea bruchid/dhora is
    (A) ***Callosobruchus affinis*** (B) *Callosobruchus maculatus*
    (C) *Callosobruchus phaseoli* (D) *Callosobruchus analis*

53. The typical damaging symptom of Pulse beetle is
    (A) Circular exit hole (B) Presence of chambers in seed
    (C) **Both A and B** (D) None of the above

54. The Family of Cigarette beetle is
    (A) **Anobiidae** (B) Tenebrionidae
    (C) Phycitidae (D) Silvanidae

55. The Family of Saw tooth grain beetle is
    (A) Anobiidae (B) Tenebrionidae
    (C) Phycitidae (D) **Silvanidae**

56. The Family of Black fungus beetle is
    (A) Anobiidae (B) **Tenebrionidae**
    (C) Phycitidae (D) Silvanidae

57. The Family of Almond moth is
    (A) Anobiidae (B) Tenebrionidae
    (C) **Phycitidae** (D) Silvanidae

58. The scientific name of Cigarette beetle is
    (A) ***Lasioderma serricorne*** (B) *Stegobium panaceum*
    (C) *Oryzaephilus surinamensis* (D) *Ephestia cautella*

59. The scientific name of Drug store beetle is
    (A) *Lasioderma serricorne* (B) ***Stegobium panaceum***
    (C) *Oryzaephilus surinamensis* (D) *Ephestia cautella*

60. The scientific name of Almond moth is
    (A) *Lasioderma serricorne* (B) *Stegobium panaceum*
    (C) *Oryzaephilus surinamensis* (D) ***Ephestia cautella***

61. The scientific name of Saw tooth grain beetle is
    (A) *Lasioderma serricorne* (B) *Stegobium panaceum*
    (C) ***Oryzaephilus surinamensis*** (D) *Ephestia cautella*

62. The Family of Angoumois grain moth is
    (A) **Gelechiidae** (B) Pyralidae
    (C) Phycitidae (D) Pyraustidae

63. The Family of Indian meal moth is
    (A) Gelechiidae (B) **Pyralidae**
    (C) Phycitidae (D) Pyraustidae

64. The Family of Rice moth is
    (A) Gelechiidae (B) **Pyralidae**
    (C) Phycitidae (D) Pyraustidae

65. The scientific name of Indian meal moth is
    (A) *Sitotroga cerealella* (B) *Corcyra cephalonica*
    (C) ***Plodia interpunctella*** (D) *Ephestia cautella*

66. The scientific name of Rice moth is
    (A) *Sitotroga cerealella* (B) ***Corcyra cephalonica***
    (C) *Plodia interpunctella* (D) *Ephestia cautella*

67. The scientific name of Angoumois grain moth is
    (A) ***Sitotroga cerealella*** (B) *orcyra cephalonica*
    (C) *Plodia interpunctella* (D) *Ephestia cautella*

68. The *Sitotroga cerealella* can be identified by
    (A) Narrow pointed wings with long hairs along posterior margin
    (B) Circular exit hole with a flap/trap door
    (C) Larva white with yellow brown head
    (D) **All of the above**

69. The hibernating stage of *Sitotroga cerealella* is
    (A) Adult only (B) **Larva only**
    (C) Both Larva and adults (D) None of the above

70. The webbing of grains indicate the infestation of

(A) *Plodia interpunctella* (B) *Corcyra cephalonica*
(C) **Both A and B** (D) None of the above

71. The silk tunnelling in the grains indicate the infestation of

(A) *Plodia interpunctella* (B) *Corcyra cephalonica*
(C) ***Ephestia cautella*** (D) None of the above

72. Clogging of mill machinery due very high level of silk tunnels in the grains is due to the infestation of

(A) *Plodia interpunctella* (B) *Corcyra cephalonica*
(C) ***Ephestia cautella*** (D) None of the above

73. The amount of food eaten by Rodents in a single day is

(A) **10% of body weight** (B) 15% of body weight
(C) 05% of body weight (D) 01% of body weight

74. The amount of food damaged by Rodents in a single day is

(A) 10 times the food they consume
(B) **20 times the food they consume**
(C) 30 times the food they consume
(D) None of the above

75. The causal organism of Rat bite fever is

(A) *Streptobacillus moniliformis* (B) *Spirillum minor*
(C) **Both A and B** (D) None of the above

76. Thescientific name of Black rat/Roof rat/House rat is

(A) ***Rattus rattus*** (B) *Mus musculus*
(C) *Bandicota bengalensis* (D) *Bandicota indica*

77. The scientific name of House mouse is

(A) *Rattus rattus* (B) ***Mus musculus***
(C) *Bandicota bengalensis* (D) *Bandicota indica*

78. The scientific name of smaller bandicot is

(A) *Rattus rattus* (B) *Mus musculus*
(C) ***Bandicota bengalensis*** (D) *Bandicota indica*

79. The scientific name of larger bandicot is

(A) *Rattus rattus* (B) *Mus musculus*
(C) *Bandicota bengalensis* (D) ***Bandicota indica***

80. The scientific name of Ship rat or Brown rat is
(A) *Rattus rattus* (B) *Mus musculus*
(C) *Bandicota bengalensis* (D) ***Rattus norvegicus***

81. The scientific name of Gebril is
(A) ***Tatera indica*** (B) *Mus platythrix*
(C) *Hystrix indica* (D) *Funambulus palmarum*

82. The Chemosteriliants for rodent control is
(A) Tepa (B) Thiotepa
(C) Apholate (D) **All of the above**

83. The scientific name of Field mice is
(A) *Tatera indica* (B) ***Mus platythrix***
(C) *Hystrix indica* (D) *Funambulus palmarum*

84. The scientific name of Squirrel is
(A) ***Tatera indica*** (B) *Mus platythrix*
(C) *Hystrix indica* (D) ***Funambulus palmarum***

85. The scientific name of Procupines is
(A) *Tatera indica* (B) *Mus platythrix*
(C) ***Hystrixindica*** (D) *Funambulus palmarum*

86. The Primary vector of Plague caused by *Yersinia pestis*, a bacterium is
(A) *Rattus rattus* (B) *R. norvegicus*
(C) *B. bengalensis* (D) ***Xenopsylla cheopis***

87. The Secondary vector of Plague caused by *Yersinia pestis*, a bacterium is
(A) *Rattus rattus* (B) *R. norvegicus*
(C) *B. bengalensis* (D) **All of the above**

88. Sweating of food grains is more under which storage conditions
(A) **Bulk storage** (B) Bag storage
(C) Air tight storage (D) All of the above

89. Heating of food grains is more under which storage conditions
(A) **Underground storage** (B) Bag storage
(C) Air tight storage (D) All of the above

90. Indian Grain Storage Institute is located at
(A) Faridabad (B) **Hapur**
(C) Ludhiana (D) New Delhi

91. Indian Grain Storage Institute (earlier GSRTC) was established in the year
   (A) **1958** (B) 1968
   (C) 1976 (D) 1988

92. Pusa bin is
   (A) Outdoor bin (B) **Indoor bin**
   (C) Both A and B (D) None of the above

93. Which of the following is the best storage structure
   (A) Gharelu Theka (B) Pusa bin
   (C) Ferro-cement bin (D) **Vaccum Process storage**

94. Which of the following is Bamboo outdoor storage structure
   (A) Galgai (B) Bukari
   (C) Topa (D) **All of the above**

95. Which of the following is Timber made outdoor storage structure
   (A) Gutibharal (B) Peti
   (C) Pathyam (D) **All of the above**

96. Which of the following is Paddy straw made outdoor storage structure
   (A) Pura (B) Puri
   (C) Morai (D) **All of the above**

97. Which of the following is Mud/brick made outdoor storage structure
   (A) Bharola (B) Pusa Bin
   (C) Kothi (D) **All of the above**

98. Which of the following is under ground storage structure
   (A) Peve (B) Khatti
   (C) Bandha (D) **All of the above**

99. The optimum moisture content of food grains for storage is
   (A) 5% (B) **10%**
   (C) 15% (D) None of the above

100. The concentration of O_2 and CO_2 in Controlled Atmosphere Storage (CAS) is
   (A) **2-4% O_2 and 9-9.50% CO_2** (B) 10-14% O_2 and 5-8% CO_2
   (C) 6-10% O_2 and 4-8% CO_2 (D) None of the above

101. The insecticides recommended as prophylactic measure in storage is
   (A) Malathion (0.50%) (B) Pyrethrum with 2% pyrethrin
   (C) Primiphos methyl (0.50%) (D) **All of the above**

102. The application rate of Malathion (0.50%) recommended as prophylactic measure in storage is

(A) 1 L/100m^3 (B) 2 L/100m^3

(C) **3 L/100m^3** (D) 5 L/100m^3

103. The application rate of Aluminium Phosphide in storage is

(A) 2 tablet/ton (B) **25 tablets/100m^3**

(C) 3 tablet/ton (D) 50 tablets/100m^3

104. The weight of tablet of Aluminium Phosphide in storage is

(A) 2 g (B) **3 g**

(C) 5 g (D) 10 g

105. The toxicity of Aluminium Phosphide is due to the

(A) Phostoxin gas (B) **Phosphine gas**

(C) Celphos gas (D) All of the above

106. The Aluminium Phosphide is sold under the trade name of

(A) Phostoxin (B) Delicia

(C) Celphos (D) **All of the above**

107. As per Insecticide Act 1968, the use of Aluminium Phosphide is under

(A) Open use category (B) **Restricted use category**

(C) Banned category (D) All of the above

108. Each Tablet ofCelphos consists of

(A) **AlP (56%) + carbamate (44%)** (B) AlP (50%) + carbamate (50%)

(C) AlP (60%) + carbamate (40%) (D) AlP (44%) + carbamate (56%)

109. The application rate of Ethylene Dibromide (EDB) in storage is

(A) **3 ml/q** (B) 5 ml/q

(C) 6 ml/q (D) None of the above

110. The which of the following is highly inflammable

(A) **Carbon tetrachloride** (B) EDB

(C) HCN (D) ALP

111. The proportion of Carbon tetrachloride mixed with Dichloro ethane is

(A) **1:3** (B) 1:5

(C) 1:10 (D) 1:20

112. The Carbon tetrachloride is mixed with Dichloro ethane to
(A) To increase shelf life (B) **Reduce fire hazard**
(C) To increase toxicity (D) To reduce toxicity

113. The acute poison which kill rats by exposure to poison bait is known as
(A) Multiple dose poison (B) **Single dose poison**
(C) Both A and (D) None of the above

114. The chronic poison which kill rats by repetitive exposure to poison bait is known as
(A) **Multiple dose poison** (B) Single dose poison
(C) Both A and B (D) None of the above

115. Which of the following is single dose poison
(A) ANTU (B) Zinc Phosphide
(C) Norbromide (D) **All of the above**

116. Which of the following is single dose poison
(A) Strychnine (B) Arsenous trioxide
(C) Barium Carbonate (D) **All of the above**

117. Which of the following is second generation single dose poison
(A) Bromadiolone (B) Vacour
(C) Silmurin (D) **All of the above**

118. Brodifacoum is a
(A) Anticoagulant (B) Single dose poison
(C) second generation acute poison (D) **All of the above**

119. Which of the following is multiple dose poison
(A) Warfarin (B) Rodofarin
(C) Racumin (D) **All of the above**

120. Which of the following is multiple dose poison
(A) Tomafarin (B) Ratafin
(C) Racumin (D) **All of the above**

121. Bromadiolone, a single dose anticoagulant is sold under the trade name of
(A) Talon (B) **Makki**
(C) RH-787 (D) None of the above

122. Brodifacoum, a single dose anticoagulant is sold under the trade name of
(A) **Talon** (B) Makki
(C) RH-787 (D) None of the above

123. Vacour, a single dose anticoagulant is sold under the trade name of

(A) Talon (B) Makki

(C) **RH-787** (D) None of the above

124. Which of the following cause Neophobia in rodents

(A) Multiple dose poison (B) **Single dose poison**

(C) Both A and (D) None of the above

125. Rodent control should be organised on

(A) Individual basis (B) **Community basis**

(C) Both A and (D) None of the above

126. The Botanical used for Rodent control

(A) **Sweet flag** (B) Sweet potato

(C) Yam (D) None of the above

127. The Sulfuryl fluoride, a fumigant is sold under the trade name of

(A) Vapourmate (B) **Profume**

(C) Ecofume (D) None of the above

128. The Ethyl formate, a fumigant is sold under the trade name of

(A) **Vapourmate** (B) Profume

(C) Ecofume (D) None of the above

129. The mixture of fumigant Phosphine @% + CO_2 98% is sold under the trade name of

(A) Vapourmate (B) Profume

(C) **Ecofume** (D) None of the above

130. Calcium Cyanide, a fumigant for rodent control is sold under the trade name of

(A) **Cyanogas** (B) Raticide

(C) Celphos (D) None of the above

131. Nor Bromide, a fumigant for rodent control is sold under the trade name of

(A) Cyanogas (B) **Raticide**

(C) Celphos (D) None of the above

132. The Chemosteriliants for rodent control is

(A) Colchicine (B) Furadantine

(C) Hempa (D) **All of the above**

Unit-12

Insect Vectors

Q 1. Match the following

1.The book "*Vectors of Plant Pathogens"* is authored by

(A) Youdeovei A & Service MW (B) Basu AN

(C) **Harris KF & Maramarosh K** (D) Maramorosch K & Harris KF

2. The book "*Pest and Vector Management in the Tropics*" is authored by

(A) **Youdeovei A & Service MW** (B) Basu AN

(C) Harris KF & Maramarosh K (D) Maramorosch K & Harris KF

3. The book " *Leafhopper Vectors and Plant Disease Agents*" is authored by

(A) Youdeovei A & Service MW (B) Basu AN

(C) Harris KF & Maramarosh K (D) **Maramorosch K & Harris KF**

4. The author of "*Crop Pest and Principal Whitefly Vector of Plant Viruses*" is

(A) Youdeovei A & Service MW (B) **Basu AN**

(C) Harris KF & Maramarosh K (D) Maramorosch K & Harris KF

5. The characteristic of Non persistent virus transmissionis

(A) No latent period (B) Low incubation period

(C) Acquisition period is very low (D) **All of the above**

6. The acquisition period in Non persistent virus transmissionis from

(A) **Seconds - minutes** (B) Minutes -hours

(C) Hours-days (D) All of the above

7. The acquisition period in Semi-persistent virus transmissionis from

(A) Seconds - minutes (B) **Minutes -hours**

(C) Hours - days (D) All of the above

8. The acquisition period in Persistent virus transmissionis from

(A) Seconds-minutes (B) **Minutes-hours**

(C) Hours-days (D) All of the above

9. Transovarial virus transmission occur in

(A) **Persistent-circulative** (B) Persistent-Propagative

(C) Semi persistent (D) All of the above

10. Vector of Stripe virus disease of Jowar is

(A) *Nilaparvata lugens* (B) *Sogatella furcifera*

(C) ***Peregrinus maidis*** (D) *Orosius albicinctus*

11. Vector of Stripe virus disease of maize is

(A) *Nilaparvata lugens* (B) *Sogatella furcifera*

(C) ***Peregrinus maidis*** (D) *Orosius albicinctus*

12. Vector of sesame phyllody is

(A) *Nilaparvata lugens* (B) *Sogatella furcifera*

(C) *Peregrinus maidis* (D) ***Orosius albicinctus***

13. Vector of sesame phyllody is

(A) *Nilaparvata lugens* (B) *Sogatella furcifera*

(C) *Peregrinus maidis* (D) ***Orosius albicinctus***

14. Vector of Coconut root wilt is

(A) ***Stephaitis typicus*** (B) *Sogatella furcifera*

(C) *Peregrinus maidis* (D) *Orosius albicinctus*

15. Fire blight of pear is transmitted by

(A) *Stephaitis typicus* (B) ***Apis mellifera***

(C) *eregrinus maidis* (D) *Orosius albicinctus*

16. Vector of Grassy stunt virus disease of rice is

(A) ***Nilaparvata lugens*** (B) *Sogatella furcifera*

(C) *Peregrinus maidis* (D) *Orosius albicinctus*

17. Vector of wheat streak virus is

(A) *Aceria cajani* (B) ***Aceria tosichella***

(C) *Schizaphis graminum* (D) *Pentalonia nigronervosa*

18. Murda disease of Chillies is transmitted by

(A) *Aceria cajani* (B) *Aceria tosichella*

(C) *Schizaphis graminum* (D) ***Polyphagotarsonemus latus***

19. Vector of Leaf curl disease of tobacco is

(A) *Aceria cajani* (B) *Aceria tosichella*

(C) *Schizaphis graminum* (D) ***Bemesia tabaci***

20. Vector of Pea leaf roll virus is
    (A) *Aceria cajani* (B) *Aceria tosichella*
    (C) ***Aphis craccivora*** (D) *Bemesia tabaci*

21. The mites suspected as vector of Mango malformation is
    (A) *Aceria cajani* (B) *Aceria tosichella*
    (C) ***Aceria mangiferae*** (D) *Bemesia tabaci*

22. Corn mosaic of maize is transmitted by
    (A) *Aceria cajani* (B) ***Pergrinus maidis***
    (C) *Schizaphis graminum* (D) *Bemesia tabaci*

23. Freckled Yellow of Sorghum is transmitted by
    (A) *Aceria cajani* (B) ***Pergrinus maidis***
    (C) *Schizaphis graminum* (D) *Bemesia tabaci*

24. Vector of Pigeon pea sterility mosaic virus is
    (A) ***Aceria cajani*** (B) *Aceria tosichella*
    (C) *Schizaphis graminum* (D) *Pentalonia nigronervosa*

25. Vector of Grassy shoot and mosaic virus in sugarcane is
    (A) *Aphis craccivora* (B) ***Rhophalosiphum maidis***
    (C) *Schizaphis graminum* (D) *Pentalonia nigronervosa*

26. Vector of Chirke virus in cardamom is
    (A) *Aphis craccivora* (B) ***Rhophalosiphum maidis***
    (C) *Schizaphis graminum* (D) *Pentalonia nigronervosa*

27. Vector of Bunchy top of bananais
    (A) *Aphis craccivora* (B) *Rhophalosiphum maidis*
    (C) *Schizaphis graminum* (D) ***Pentalonia nigronervosa***

28. Vector of Leaf roll and Y viruses in potato is
    (A) *Aphis craccivora* (B) ***Aphis gossypii***
    (C) *Schizaphis graminum* (D) *Pentalonia nigronervosa*

29. Vector of Common broad bean mosaic virus in beans is
    (A) ***Aphis craccivora*** (B) *Aphisgossypii*
    (C) *Schizaphis graminum* (D) *Pentalonia nigronervosa*

30. TheAnthrax is vectored by
    (A) Robber flies (B) **Horse flies**
    (C) Syrphid flies (D) Gnats

31. The Banana aphid, *Rhophalosiphum maidis* is vector of
    - (A) Cardamom Mosaic virus
    - (B) Bunchy top of banana
    - (C) Katte virus Cardamom
    - (D) **All of the above**

32. The vector of Leaf roll and 'Y' viruses in potato is
    - (A) *Aphis craccivora*
    - (B) ***Aphis gossypii***
    - (C) *Schizaphis graminum*
    - (D) *Pentalonia nigronervosa*

33. The vector of common broad bean mosaic virus is
    - (A) ***Aphis craccivora***
    - (B) *Aphis gossypii*
    - (C) *Schizaphis graminum*
    - (D) *Pentalonia nigronervosa*

34. Vector of Rice Yellow disease
    - (A) *Nilaparvata lugens*
    - (B) *Nephotettix viriscens*
    - (C) ***Sogatella furcifera***
    - (D) *Recilia dorsalis*

35. Vector of Stripe disease of Jowar and Maize
    - (A) *Nilaparvata lugens*
    - (B) *Nephotettix viriscens*
    - (C) *Sogatella furcifera*
    - (D) ***Peregrinus maidis***

36. Rice green leaf hopper, *Nephotettix viriscens* is vector of
    - (A) Yellow dwarf virus
    - (B) Tungro virus
    - (C) Dwarf virus
    - (D) **All of the above**

37. Vector of little leaf disease of brinjal is
    - (A) ***Cestius phycitis***
    - (B) *Orosius albicinctus*
    - (C) *Peregrinus maidis*
    - (D) *Idioscopus niveopsparsus*

38. Rice green leaf hopper, *Nephotettix viriscens* is vector of
    - (A) Yellow dwarf virus
    - (B) Tungro virus
    - (C) Dwarf virus
    - (D) **All of the above**

39. The vector of cotton leaf curl virus is
    - (A) ***Bemesia tabaci***
    - (B) *Trialeurodes ricini*
    - (C) *Trialeurodes dispersus*
    - (D) *Aleurolobus barodensis*

40. The vector of leaf curl virus of chillies is
    - (A) *Bemesia tabaci*
    - (B) *Trialeurodes ricini*
    - (C) ***Scirtothrips dorsalis***
    - (D) *Aleurolobus barodensis*

41. The tomato spotted wilt virus is transmitted by
    - (A) *Bemesia tabaci*
    - (B) *Trialeurodes ricini*
    - (C) *Trialeurodes dispersus*
    - (D) ***Frankiniella occidentalis***

42. The vector of Vein clearing disease of okra is
(A) ***Bemesia tabaci*** (B) *Trialeurodes ricini*
(C) *Trialeurodes dispersus* (D) *Aleurolobus barodensis*

43. The vector of Sesamum leaf curl virus is
(A) ***Bemesia tabaci*** (B) *Trialeurodes ricini*
(C) *Trialeurodes dispersus* (D) *Aleurolobus barodensis*

44. The vector of mungbean yellow mosaic virus is
(A) Bean thrips (B) **Whitefly**
(C) Bean aphid (D) Bean Jassid

45. The vector of Pea leaf roll virus is
(A) Bean thrips (B) Whitefly
(C) **Bean aphid** (D) Bean Jassid

46. The bean aphid, *Aphis craccivora* is a vector of
(A) Cucumber mosaic virus (B) Pea leaf roll virus
(C) Tobacco streak virus (D) **All of the above**

47. Mottling virus diseases in sugarcane is transmitted by
(A) *Aphis craccivora* (B) ***Saccharicoccus sacchari***
(C) *Aproaerma modicella* (D) *Frankliniella schultzei*

48. The vector of Peanut stripe virus diseases is
(A) ***Aphis craccivora*** (B) *Empoasca kerri*
(C) *Aproaerma modicella* (D) *Frankliniella schultzei*

49. The vector of Peanut bud Necrosis virus diseases is
(A) *Aphis craccivora* (B) *Empoasca kerri*
(C) *Aproaerma modicella* (D) ***Frankliniella schultzei***

50. The vector of Orange leaf virus disease is
(A) *Nephotettix viriscens* (B) *Sogatella furcifera*
(C) *Nilaparvata lugens* (D) ***Recilia dorsalis***

51. The Tungro virus is transmitted by
(A) ***Nephotettix viriscens*** (B) *Sogatella furcifera*
(C) *Nilaparvata lugens* (D) *Recilia dorsalis*

52. The Grassy stunt virus is transmitted by
(A) *Nephotettix viriscens* (B) *Sogatella furcifera*
(C) ***Nilaparvata lugens*** (D) *Recilia dorsalis*

53. The vector of mycoplasma Yellow dwarf disease is
(A) *Nephotettix viriscens* (B) *Sogatella furcifera*
(C) *Nilaparvata lugens* (D) ***Recilia dorsalis***

54. The Insect vector of cardamom disease/Katte disease is
(A) *Aspidiotus destructor* (B) ***Pentalonia nigreonervosa***
(C) *Aphis craccivora* (D) *Scirtothrips cardamomi*

55. Rossette virus disease of Groundnut is transmitted by
(A) *Aspidiotus destructor* (B) *Pentalonia nigreonervosa*
(C) ***Aphis craccivora*** (D) *Scirtothrips cardamomi*

56. The Insect vector of *Chirke* disease of cardamom is
(A) *Aspidiotus destructor* (B) *Pentalonia nigreonervosa*
(C) *Aphis craccivora* (D) ***Rhophalosiphum maidis***

57. The Insect vector of *Amomum*Mosiac or Foorky disease of large cardamom is
(A) *Aspidiotus destructor* (B) ***Pentalonia nigreonervosa***
(C) *Aphis craccivora* (D) *Scirtothrips cardamomi*

58. The Vector of citrus greening disease is
(A) ***Diaphirina citri*** (B) *Toxoptera citricidi*
(C) *Phyllocnistis citrella* (D) *Aleurocanthus woglumi*

59. The vector of bunchy top disease in banana is
(A) *Odioporus longicollis* (B) *Cosmpolites sordidus*
(C) *Nacoleia octasema* (D) ***Pentalonia nigeronervosa***

60. The Onion thrips, *Thrips tabaci* is a vector of
(A) Peas Streak virus (B) Pineapple yellow spot virus
(C) **Both A and B** (D) None of the above

61. The vector of chillies leaf virus is
(A) *Scirtothrips dorsalis* (B) ***Bemesia tabaci***
(C) *Polyphagotarsonemus latus* (D) *Aleurolobus barodensis*

62. The vector of Kala Azar is
(A) *Cimex lectularius* (B) *Cimex hemipterus*
(C) **Both A and B** (D) None of the above

63. Vector of *Trypanosoma cruzi* causing Trypanosomiasis is
(A) *Tritoma* sp. (B) *Rhodinus prolixus*
(C) Both A and B (D) None of the above

64. The Primary vector of Plague caused by *Yersinia pestis*, a bacterium is

(A) *Rattus rattus* (B) *R. norvegicus*

(C) *B. bengalensis* (D) ***Xenopsylla cheopis***

65. The Secondary vector of Plague caused by *Yersinia pestis*, a bacterium is

(A) *Rattus rattus* (B) *R. norvegicus*

(C) *B. bengalensis* (D) **All of the above**

Unit-13

Beneficial Insects

Q 1. Match the following

1. The author of "Beekeeping in India" is
   (A) **Sardar Singh** (B) DP Abrol
   (C) AS Atwal (D) H Aruga

2. The author of "Bees and Beekeeping in India" is
   (A) Sardar Singh (B) **DP Abrol**
   (C) AS Atwal (D) H Aruga

3. The scientific name of tropical Tasar silkworm is
   (A) ***Antheraea mylitta*** (B) *Antheraea proylei*
   (C) *Antheraea roylei* (D) *Antheraea perenii*

4. The scientific name of Oak Tasar silkworm is
   (A) *Antheraea roylei* (B) *Antheraea proylei*
   (C) *Antheraea perenii* (D) **All of the above**

5. The scientific name of Eri silkworm is
   (A) ***Samia cynthia*** (B) *Antheraea proylei*
   (C) *Antheraea assama* (D) All of the above

6. The scientific name of Muga silkworm is
   (A) *Samia cynthia* (B) *Antheraea proylei*
   (C) ***Antheraea assama*** (D) All of the above

7. The family of Tasar silkworm is
   (A) Bombycidae (B) **Saturniidae**
   (C) Antheridae (D) All of the above

8. The family of Muga silkworm is
   (A) Bombycidae (B) **Saturniidae**
   (C) Antheridae (D) All of the above

9. The Host plants of Muga silkworm is

(A) Som (*Persia bombycina*) (B) Sualu (*Litsea polyantha*)

(C) **Both A and B** (D) None of the above

10. The Host plants of Tropical Tasar silkworm (*A. mylitta*) is

(A) Asan (*Terminalia tomentosa*) (B) Sal (*Shorea robusta*)

(C) Arjun (*Terminalia arjuna*) (D) **All of the above**

11. Mulberry leaves picked from shoots after drastic pruning have

(A) Soft leaf (B) Coarse leaf

(C) **Leaf with less water content** (D) More fibre

12. Mulberry leaves picked from shoots after drastic pruning have

(A) Soft leaf (B) Coarse leaf

(C) **Leaf with less water content** (D) More fibre

13. Ideal chawki rearing is done by

(A) Using paraffin paper (B) Rearing in boxes with lids

(C) Rearing in boxes without lids (D) **All of them**

14. The floss percentage of cocoon in bivoltine races is

(A) 1% (B) **2%**

(C) 3% (D) 4%

15. The moisture (water) in an ideal honey on an average is

(A) **17.2%** (B) 18.2%

(C) 19.2% (D) 21%16. Pebrine disease of silkworm is caused by

(A) *Streptococci* (B) ***Nosema bombycis***

(C) *Bacillus pyocyaneus* (D) *Staphylococci*

17. Sotto disease in silkworm is caused by

(A) *Bacillus prodigiosus* (B) *Proteus*

(B) *Serratia marcescens* (D) ***Bacillus thuringiensis***

18. Grasserie disease in silkworm is caused by

(A) **NPV** (B) Bacteria

(C) Sporozoa (D) Fungus

19. Flacherie disease in silkworm is caused by

(A) NPV (B) **Bacteria**

(C) Sporozoa (D) Fungus

20. The Bacteria causing Flacherie disease in silkworm is
    (A) ***Bacillus bombysepticus*** (B) *Nosema bombycis*
    (C) *Streptococcus bacillus* (D) *Streptobacillus bombycis*
21. Nuclear polyhedrosis disease in silkworm is caused by
    (A) ***Borrelina virus*** (B) *Smithia virus*
    (C) *Morator virus* (D) None of them
22. White muscardine disease in silkworm is caused by
    (A) *Metarrhizium anisopliae* (B) ***Spicaria prasina***
    (C) *Beuveria bassiana* (D) *Isaria fasinosa*
23. Green muscardine disease in silkworm is caused by
    (A) *Metarrhizium anisopliae* (B) *Spicaria prasina*
    (C) ***Beuveria bassiana*** (D) *Isaria fasinosa*
24. Yellow muscardine disease in silkworm is caused by
    (A) *Metarrhizium anisopliae* (B) *Spicaria prasina*
    (C) *Beuveria bassiana* (D) ***Isaria fasinosa***
25. The scientific name of Uzi fly is
    (A) *Holocera pulvicera* (B) ***Exorista sorbillans***
    (C) *Dermestes cadaverinus* (D) *Galleria mellonella*
26. Headquarter of Central Silk Board is located at
    (A) **Bangalore (Karnataka)** (B) Ranchi (Jharkhand)
    (C) Hosur (Tamil Nadu) (D) Lahdoigarh (Assam)
27. Headquarter of Central Sericultural Germplasm Resource Centre is located at
    (A) Bangalore (Karnataka) (B) Ranchi (Jharkhand)
    (C) **Hosur (Tamil Nadu)** (D) Lahdoigarh (Assam)
28. Headquarter of Central Tasar Research and Training Institute is located at
    (A) Bangalore (Karnataka) (B) **Ranchi (Jharkhand)**
    (C) Hosur (Tamil Nadu) (D) Lahdoigarh (Assam)
29. Headquarter of Central Muga Eri Research and Training Institute is located at
    (A) Bangalore (Karnataka) (B) Ranchi (Jharkhand)
    (C) Hosur (Tamil Nadu) (D) **Lahdoigarh (Assam)**
30. The Egg diapause is present in
    (A) Muga Silkworm (B) Eri Silkworm
    (C) Tasar Silkworm (D) **Mulberry silkworm**

31. The Pupal diapause is present in
    (A) Muga Silkworm (B) Eri Silkworm
    (C) **Tasar Silkworm** (D) Mulberry silkworm

32. The Poor man's silk is
    (A) Muga Silkworm (B) **Eri Silkworm**
    (C) Tasar Silkworm (D) Mulberry Silkworm

33. The Golden silk is
    (A) **Muga Silkworm** (B) Eri Silkworm
    (C) Tasar Silkworm (D) Mulberry Silkworm

34. Ribosomes are composed of
    (A) DNA & Proteins (B) DNA only
    (C) DNA & RNA (D) **RNA & Proteins**

35. Which of the following is called **'suicide bag'**
    (A) **Lysosomes** (B) Peroxisomes
    (C) Centrosome (D) Glyoxisome

36. Ribosomes are centres of
    (A) Respiration (B) **Protein synthesis**
    (C) Photosynthesis (D) Lipid synthesis

37. The organelles called **'power house of cell'** are
    (A) **Mitochondria** (B) Golgi apparatus
    (C) Ribosomes (D) Lysosomes

38. Four bases which form the code words for DNA language are:
    (A) ACTU (B) UCTA
    (C) AGTU (D) **ATCG**

39. Plant cell wall possesses
    (A) Cellulose (B) Hemicellulos
    (C) Pectin (D) **All of these**

40. Which of the following is living
    (A) Cellulose (B) Nucleus
    (C) Cytoplasm (D) **All of these**

41. Silk protein is derived from
    (A) Soil (B) **Silkworm**
    (C) Host plant (D) None of them

42. Both quality and quantity of leaf harvest is obtained from:
(A) Bottom pruning (B) Midcut pruning
(C) Top Clipping (D) **All types of pruning**

43. Irrigation coupled with fertilizater application increases leaf yield by
(A) **3-5 times** (B) 4-6 times
(C) 7-8 times (D) 8-10 times

44. For ideal mulberry cultivation, soil should be
(A) Alluvial (B) Volcanic
(C) **Both Alluvial & volcanic** (D) None of them

45. Mulberry silkworm is monophagous because of
(A) Host specificity (B) Nutritional requirement
(C) Green leaves (D) **Presence of attractants**

46. Crude proteins in mulberry leaves can be increased by
(A) Phosphorus (B) **Nitrogen**
(C) Potassium (D) None of them

47. Reclamation of alkaline soil for mulberry can be done by
(A) Lime (B) **Gypsum**
(C) Press mud (D) None of them

48. Wilting of tender mulberry leaves is due to the deficiency of
(A) Zinc (B) Manganese
(C) **Copper** (D) Iron

49. Pruning of mulberry plant results into
(A) Devitalizing (B) Shaping of plant
(C) **Improving yield** (D) To have more leaf harvest for year

50. Mulberry is native of
(A) India (B) Japan
(C) Korea (D) **China**

51. Silk Industry originated from:
(A) Kotan (B) **Chang-tong**
(C) Tibet (D) Japan

52. Mulberry grows in
(A) Mediterrarean Climate (B) Extreme Variation in climate
(C) Mansoon type of climate (D) **All of them**

53. For optimum growth of mulberry temperature range is:
(A) 10-30°C (B) **13-38°C**
(C) 15-38°C (D) 20-38°C

54. The number of species belonging to genus *Morus* are
(A) < 10 (B) > 15
(C) > 20 (D) < 20

55. Mulberry plant is
(A) Perennial (B) Deciduous
(C) **Both A & B** (D) None of them

56. Dormancy of mulberry bud is due to
(A) **Low temperature** (B) Nitrogen deficiency
(C) Auxins (D) Photoperiodism

57. Mulberry seed remains viable up to
(A) Few days (B) **Few weeks**
(C) Two months (D) Three months

58. Cuttings having high content of any of the following root readily
(A) Nitrogen (B) **Carbohydrates**
(C) Protein (D) Amino acid

59. Which season leaf is better for silkworm rearing
(A) **Spring Season** (B) Summer Season
(C) Autumn Season (D) Rainy Season

60. Best method of raising mulberry is
(A) **Graft** (B) Seed
(C) Vegetatively (D) Layering

61. For producing one gram of dry matter the water requirement of mulberry plant should be
(A) 250-350 ml (B) 260-360 ml
(C) 270-385 ml (D) **280-400 ml**

62. Which fertilizer is most important and vital for leaf productivity
(A) Phosphorus (B) **Nitrogen**
(C) Potassium (D) None of them

63. Name most common organic manure used in mulberry
(A) **Farmyard Manure** (B) Sun-hemp
(C) Cowpeas (D) Neem Cake

64. Silkworm *Bombyx mori* moth belongs to

(A) Apterygota (B) Exopterygota

(C) **Endopterygota** (D) None of them

65. Larger and heavier eggs of silkworm *Bombyx mori* belongs to

(A) Indian races (B) Japanese races

(C) Chinese races (D) **European races**

66. Silkworm head comprises of

(A) Four segments (B) Five segments

(C) **Six segments** (D) Eight segments

67. *Bombyx mori* have

(A) 2 pairs of ocelli (B) 4 pairs of ocelli

(C) **6 pairs of ocelli** (D) 8 pairs of ocelli

68. Silkworm *Bombyx mori* discriminate the host plant leaves by

(A) Maxillary lobe (B) Maxillary palpi

(C) **Both A&B** (D) None of them

69. Silkworm expels silk from

(A) Labium (B) Labrum

(C) Mandible (D) **Spinneret**

70. The number of true legs in silkworm are

(A) Two (B) Three

(C) **Six** (D) No legs

71. Claws are used for:

(A) Crawling (B) Climbing

(C) Defence (D) **Holding**

72. Abdomen of silkworm comprises of

(A) 8 segments (B) 9 Segments

(C) **10 Segments** (D) 11 Segments

73. Silkworm larva bears a caudal horn on

(A) 6th segment (B) 7th segment

(C) **8th segment** (D) None of them

74. Sexual marking of silkworm larvae can be seen from ventral side on

(A) 6th& 7th abdominal segment (B) 7th& 8th abdominal segment

(C) **8th& 9th abdominal segment** (D) 9th& 10th abdominal segment

75. The spiracle number in silkworm are
   (A) **9 pairs** (B) 8 pairs
   (C) 6 pairs (D) None of them

76. Mulberry silk moth is
   (A) Fast flier (B) **Incapable to fly**
   (C) Can fly moderately (D) Has no wings

77. Mulberry silkworm moth has
   (A) Ocelli (B) Simple eyes
   (C) **Compound eyes** (D) No eyes

78. Mulberry silkworm moth has
   (A) **One pair of small antennae** (B) Single antennae
   (C) No antennae (D) Vestigeal antennae

79. The number of spiracles in silkworm moth are
   (A) **6 pairs** (B) 7 pairs
   (C) 8 pairs (D) 9 pairs

80. The digestive fluid in silkworm is secreted by
   (A) Epithelium cells (B) **Goblet cells**
   (C) Cylindrical cells (D) None of them

81. Probosis of silkworm adult moth is
   (A) Well developed (B) Vestigial
   (C) Scantly developed (D) **Non functional**

82. The number of Malpighian tubes in adult silkworm are
   (A) Three (B) Four
   (C) Five (D) **Six**

83. The brain of silkworm comprises of
   (A) One lobe (B) **Two lobules**
   (C) Three lobules (D) None of them

84. The ventral nerve cord in silkworm is formed by
   (A) One thoracic ganglia (B) Two thoracic ganglia
   (C) **Three thoracic ganglia** (D) A series of thoracic ganglia

85. In silkworm larva a pair of ovaries originates form
   (A) 4th abdominal segment (B) 5th abdominal segment
   (C) **6th abdominal segment** (D) 8th abdominal segment

86. In silkworm larva a pair of testes are situated at

(A) 4th abdominal segment (B) **5th abdominal segment**

(C) 6th abdominal segment (D) 8th abdominal segment

87. The number of ovarioles in silkworm ovary are

(A) One (B) Two

(C) Three (D) **Four**

88. The adhesive material which coats the silkworm egg is produced by

(A) Bursa copulatrix (B) Aedeagus

(C) **Female Accessory glands** (D) Spermatheca

89. Silk components; Fibrion and Sericin are secreted in

(A) 1 & 2 larval age (B) 2 & 3 larval age

(C) 3 & 4 larval age (D) **4 & 5 larval age**

90. To complete its cycle, silkworm passes through

(A) Two stages (B) Three stages

(C) **Four stages** (D) Five stages

91. Mulberry silkworm moults

(A) Once only (B) Twice

(C) Thrice (D) **Four times**

92. The optimum temperature range for preservation of seed cocoons is

(A) 20-22°C (B) **23-25°C**

(C) 26-28°C (D) 28-30°C

93. Optimum humidity for preservation of seed cocoon is

(A) **65-70%** (B) 70-75%

(C) 75-80% (D) 80-85%

94. Ideal coupling time duration for obtaining maximum viable silkworm eggs is

(A) **1-2 hrs.** (B) 3-4 hrs.

(C) 5-6 hrs (D) 6 hrs.

95. Optimum temperature & humidity for incubating silkworm eggs is

(A) 20-22°C & 70-75% RH (B) 22-24°C & 75-80% RH

(C) **24-25°C & 80-85% RH** (D) 25-26°C & 75-80% RH

96. Commonly used disinfectants in silkworm rearing is

(A) Lime (B) Formaldehyde

(C) Bleaching powder (D) **All of the these**

97. Loose silkworm egg brushing is carried out by

(A) A feather (B) **Using a mosquito net**

(C) Tapping of newly born worms (D) None of the them

98. Silkworm are

(A) Homothermal (B) **Poikilothermal**

(C) No blood (D) None of them

99. Mulberry seed germinates better in

(A) **Bright light** (B) Defused light

(C) Darkness (D) None of the three

100. The optimum range for germination of mulberry seed is

(A) 18-21ºC (B) 27-30ºC

(C) **33-36ºC** (D) 36-3ºC

101. Spacing between plants for rainfed mulberry

(A) 0.6-0.48m (B) 0.7-0.49m

(C) 0.8-0.50m (D) **0.9-0.45m**

102. For propagation of mulberry cutting, preference is given to

(A) Young seedlings (B) Old seedlings

(C) **Mature seedlings** (D) None of them

103. For propagation of mulberry through grafting the stock should be

(A) **1-2 year old seedling** (B) 2-3 year old seedling

(C) 3-4 year old seedling (D) Age of plant does not matter

104. Crown grafting in mulberry is done by

(A) Inserting 1 scions (B) Inserting 2 scions

(C) Inserting 3 scions (D) **Inserting more than one scions**

105. For root grafting in mulberry the thickness of root should be

(A) 0.5 to 3 cm (B) **0.6 to 2.5 cm**

(C) 0.8 to 2.7 cm (D) 1 to 3 cm

106. Ideal atmospheric humidity for mulberry growth is

(A) 50-65% (B) 55-75%

(C) **65-80%** (D) 75-80%

107. Inter-row spacing in small mulberry trees should be

(A) 1.5-2.0m (B) 2.0-2.5m

(C) **2.5-3m** (D) None of them

108. Inter plant spacing in small mulberry trees should be
(A) 0.6-0.8m (B) **0.8-1.2m**
(C) 1.2-2.0m (D) None of them

109. Spacing for rain fed mulberry (pit system) between rows should be
(A) 0.7-0.85m (B) 0.8-0.95m
(C) **0.9-0.75m** (D) 1-0.85m

110. Spacing for irrigated mulberry between rows should be
(A) 0.1-0.40m (B) 0.2-0.45m
(C) **0.3-0.45m** (D) 0.45-0.60m

111. Spacing between plants for irrigated mulberry
(A) **0.45-0.60m** (B) 0.40-0.50m
(C) 0.35-0.45m (D) None of them

112. World production of mulberry silk is
(A) 80% (B) 85%
(C) 90% (D) **95%**

113. The physical basis of life is
(A) Nucleus (B) Cell
(C) **Food** (D) Protoplasm

114. Malpighian tubes in silkworms are organs of
(A) Respiration (B) **Excretion**
(C) Circulation (D) None of them.

115. The tracheae of silkworm are similar in structure to that of:
(A) Integument (B) **Cuticle**
(C) Epidermis (D) None of them

116. The head movement of silkworm larvae are controlled by
(A) **Cervical muscles** (B) Muscles of mouth parts
(C) Pleural muscles (D) Dorso-ventral muscles

117. The Kusumi strain of Lac insect is reared on
(A) *Zizyphus mauritiana* (B) *Butea monosperma*
(C) *Shorea robusta* (D) ***Schleichera oleon***

118. The Rangeeni strain of Lac insect is reared on
(A) *Zizyphus mauritiana* (B) *Butea monosperma*
(C) *Shorea robusta* (D) **All of the above**

119. The Kusumi strain of Lac insect is

(A) Jethwi (B) Aghani

(C) Katki (D) **Both A and B**

120. The Rangeeni strain of Lac insect is

(A) Katki (B) Baisakhi

(C) Jethwi (D) **Both A and B**

121. The superior quality lac is produced from which strain of Lac insect

(A) Rangeeni (B) **Kusumi**

(C) Jethwi (D) Baisakhi

122. Lac glands are present on all body parts except

(A) Mouth parts (B) Two breathing pores

(C) Anus (D) **All of the above**

123. Lac on the twigs after nymphs escaped from the brood lac is

(A) **Stick lac/Phunki** (B) Seed Lac

(C) Kiri (D) Shellac

124. Scrapped Lac crushed and washed in water to remove impurities is

(A) Stick lac/Phunki (B) **Seed Lac/Chowri**

(C) Kiri (D) Shellac

125. Impurities left in bag after heating lac for extraction is

(A) Sticklac (B) Seed Lac

(C) **Kiri** (D) Shellac

126. Pure lac stretched in to thin small flakes is

(A) Sticklac (B) Seed Lac

(C) Kiri (D) **Shellac**

127. Honeybee queen bee as a female that laid eggs was first reported by

(A) **Luis Méndez de Torres** (B) Lorenzo Langstroth

(C) Richard Remnant (D) Karl Von Frisch

128. Honeybee worker as a imperfect female was first reported by

(A) Luis Méndez de Torres (B) Lorenzo Langstroth

(C) **RichardRemnant** (D) Karl Von Frisch

129. Dance Language of honeybee was first reported by

(A) Luis Méndez de Torres (B) Lorenzo Langstroth

(C) Richard Remnant (D) **Karl Von Frisch**

130. Bee hive with movable frames was first invented by
(A) Luis Méndez de Torres (B) **Lorenzo Langstroth**
(C) Richard Remnant (D) Karl Von Frisch

131. Father of beekeeping is
(A) Luis Méndez de Torres (B) **Lorenzo Langstroth**
(C) Richard Remnant (D) Karl Von Frisch

132. Father of beekeeping in India is
(A) R.N. Mutto (B) Lorenzo Langstroth
(C) **A.S Atwal** (D) Revd. Newton

133. Utility of honeybee as pollinator was first demonstrated by
(A) Luis Méndez de Torres (B) Lorenzo Langstroth
(C) **J B Free** (D) Karl Von Frisch

134. Central Bee Research and Training Institute is located at
(A) Ludhiana (B) **Pune**
(C) Bangalore (D) New Delhi

135. The Indian honeybee is
(A) ***Apis cerana*** (B) *Apis mellifera*
(C) *Apis dorsata* (D) *Apis florea*

136. The Rock/Giant honeybee is
(A) *Apis cerana* (B) *Apis mellifera*
(C) ***Apis dorsata*** (D) *Apis florea*

137. The Italian / Golden honeybee is
(A) *Apis cerana* (B) ***Apis mellifera***
(C) *Apis dorsata* (D) *Apis florea*

138. The Little honeybee is
(A) *Apis cerana* (B) *Apis mellifera*
(C) *Apis dorsata* (D) ***Apis florea***

139. The honeybee species whose nests are single vertical comb
(A) *Apis laboriosa* (B) *Apis florea*
(C) *Apis dorsata* (D) **All of the above**

140. The honeybee species who make parallel combs
(A) *Apis cerana* (B) *Apis mellifera*
(C) *Apis koschenikovi* (D) **All of the above**

141. The Cliff honeybee is

(A) ***Apis laboriosa*** (B) *Apis florea*

(C) *Apis dorsata* (D) All of the above

142. The open nesting honeybee species is

(A) *Apis laboriosa* (B) *Apis florea*

(C) *Apis dorsata* (D) **All of the above**

143. The Closed nesting honeybee species

(A) *Apis cerana* (B) *Apis mellifera*

(C) *Apis koschenikovi* (D) **All of the above**

144. The most aggressive honeybee species

(A) *Apis cerana* (B) *Apis mellifera*

(C) ***Apis dorsata*** (D) *Apis florea*

145. The domesticated honeybee species is

(A) *Apis cerana* (B) *Apis mellifera*

(C) **Both A and B** (D) *Apis florea*

146. The pollen load (mg) carried by worker bee of *Apis mellifera* is

(A) 7-14 mg (B) **12-29 mg**

(C) 30-45 mg (D) 10-20 mg

147. The pollen load (mg) carried by worker bee of *Apis cerana* is

(A) **7-14 mg** (B) 12-29 mg

(C) 30-45 mg (D) 10-20 mg

148. The average number of eggs laid by queen bee of *Apis cerana* is

(A) 200-400/day (B) **500-800/day**

(C) 1000-1500/day (D) 1500-2000/day

149. The average number of eggs laid by queen bee of *Apis cerana* is

(A) 200-400/day (B) 800-1000/day

(C) **1000-1500/day** (D) 1500-2000/day

150. The royal jelly is secreted from which glands

(A) Mandibular glands (B) **Pharyngeal glands**

(C) Wax glands (D) None of the above

151. The Queen substance is secreted from which glands

(A) **Mandibular glands** (B) Pharyngeal glands

(C) Wax glands (D) None of the above

152. The Queen substance is
   (A) **9-ODA** (B) Decenoicacid
   (C) OxiDecenoic Acid (D) None of the above

153. The replacement of aged queen with a new one is
   (A) **Replacement** (B) **Supercedure**
   (C) Carry over (D) None of the above

154. The Function of Queen substance is
   (A) Prevent swarming and absconding of colonies
   (B) Prevent development of ovary in workers
   (C) Maintain Colony cohesion
   (D) **All of the above**

155. The development duration of queen egg to emergence is
   (A) **16 days** (B) 21 days
   (C) 24 days (D) 58 days

156. The development duration of Worker bee from egg to emergence is
   (A) 16days (B) **21 days**
   (C) 24 days (D) 58 days

157. The development duration of Drone bee egg to emergence is
   (A) 16 days (B) 21 days
   (C) **24 days** (D) 58 days

158. The pollen basket is present in
   (A) **Worker bee** (B) Drones
   (C) Queen bee (D) All of the above

159. The colony multiplication and migration during favourable conditions is
   (A) Migration (B) Absconding
   (C) **Swarming** (D) All of the above

160. The colony desertion of colony during unfavourable conditions is
   (A) Migration (B) **Absconding**
   (C) Swarming (D) All of the above

161. During swarming
   (A) Whole colony migrate (B) **Part of colony migrate**
   (C) Both A and B (D) None of the above

162. During Absconding

(A) **Whole colony migrate** (B) Part of colony migrate

(C) Both A and B (D) None of the above

163. The tendency of stealing in bee colony is known as

(A) Migration (B) Absconding

(C) Swarming (D) **Robbing**

164. Worker bee perform round dance when the distance from the food source is

(A) **<50m** (B) 50-100m

(C) >100 (D) None of the above

165. Worker bee perform Wag tail dance when the distance from the food source is

(A) <50m (B) 50-100m

(C) **>100** (D) None of the above

166. The bee space required by *Apis mellifera* for normal movement and functioning in between two adjacent combs is

(A) 7-9mm (B) **10mm**

(C) 12-15mm (D) 15-18mm

167. The bee space required by *Apis cerana* for normal movement and functioning in between two adjacent combs is

(A) **7-9mm** (B) 10mm

(C) 12-15mm (D) 15-18mm

168. Which one is more beneficial and high return yielding

(A) Stationary beekeeping (B) **Migratory beekeeping**

(C) Traditional beekeeping (D) Wild bee keeping

169. The specific gravity of pure honey ranges between

(A) **1.25-1.44** (B) 1.44-1.65

(C) 1.10-1.15 (D) 1.67-1.98

170. The granulation of honey is due to

(A) **Dextrose** (B) Laevulose

(C) Glucose (D) Fructose

171. High ratio of Levulose/Dextrose indicates

(A) High granulation (B) **Less granulation**

(C) Both A and B (D) None of the above

172. Wax glands in worker bees are present in

(A) 4-7th **abdominal sternite** (B) 5-8th abdominal sternite

(C) 3-7th abdominal sternite (D) None of the above

173. Pollen basket in worker bees are present on

(A) **Tibia of Forelegs** (B) Tibia of middle leg

(C) Tibia of hind leg (D) None of the above

174. The mouthpart of worker bees are

(A) **Chewing and Lapping type** (B) Biting and Chewing type

(C) Piercing and sucking type (D) Siphoning type

175. The Antennae of worker bees are

(A) Moniliform (B) **Geniculate**

(C) Aristate (D) Stylate

176. Honey stomach in worker bee is a modified

(A) Ventriculus (B) **Crop**

(C) Pharynx (D) Oesophagous

177. The *Apis mellifera* species domesticated and commercially used in India is

(A) ***Apis mellifera liguistica*** (B) *Apis melllifera mellifera*

(C) *Apis melllifera carnica* (D) *Apis melllifera major*

178. The American Foul brood is caused by

(A) *Melissococcus plutonis* (B) ***Paenibacillus larvae***

(C) *Morator aetatulus* (D) *Nosema apis*

179. The European Foul brood is caused by

(A) ***Melissococcus plutonis*** (B) *Bacillus larvae*

(C) *Morator aetatulus* (D) *Nosema apis*

180. The dysentery disease in honeybee is caused by

(A) *Melissococcus plutonis* (B) *Bacillus larvae*

(C) *Morator aetatulus* (D) ***Nosema apis***

181. The Sac brood disease in honeybee is caused by

(A) *Melissococcus plutonis* (B) *Bacillus larvae*

(C) ***Morator aetatulus*** (D) *Nosema apis*

182. The stone brood disease in honeybee is caused by

(A) *Ascosphaera apis* (B) ***Aspergillus flavus***

(C) *Morator aetatulus* (D) *Nosema apis*

183. The Chalk brood disease in honeybee is caused by
    (A) ***Ascosphaera apis*** (B) *Aspergillus flavus*
    (C) *Morator aetatulus* (D) *Nosema apis*

184. The Lesser wax moth in honeybee colonies is
    (A) *Galleria melonella* (B) ***Achroia grisella***
    (C) *Vespa velutina* (D) *Vespa cincta*

185. The Greater wax moth in honeybee colonies is
    (A) ***Galleria melonella*** (B) *Achroia grisella*
    (C) *Vespa velutina* (D) *Vespa cincta*

186. The Giant wasp predating honeybee colonies is
    (A) *Galleria melonella* (B) *Achroia grisella*
    (C) ***Vespa velutina*** (D) *Vespa cincta*

187. The Yellow banded wasp predating honeybee colonies is
    (A) *Galleria melonella* (B) *Achroia grisella*
    (C) *Vespa velutina* (D) ***Vespacincta***

188. Isle of Wight disease in honeybee colonies is caused by
    (A) ***Acarpis woodi*** (B) *Varroa destructor*
    (C) *Varroa jacobsoni* (D) All of the above

189. Tracheal mite in honeybee colonies is
    (A) ***Acarpis woodi*** (B) *Varroa destructor*
    (C) *Varroa jacobsoni* (D) All of the above

Unit-14

Plant Nematology

Q 1. Match the following

1. The author of "*Plant Nematology*" is

(A) RK Walia and HK Bajaj (B) G Thorne

(C) **RN Perry and M Moens** (D) DS Bhatti and RK Walia

2. The author of "*Principles of Nematology*" is

(A) RK Walia and HK Bajaj (B) **G Thorne**

(C) RN Perry and M moens (D) DS Bhatti and RK Walia

3. The author of "*Textbook on Introductory Plant Nematology*" is

(A) **RK Walia and HK Bajaj** (B) G Thorne

(C) RN Perry and M moens (D) DS Bhatti and RK Walia

4. The author of "*Nematode Pests of Crops*" is

(A) RK Walia and HK Bajaj (B) G Thorne

(C) RN Perry and M moens (D) **DS Bhatti and RK Walia**

5. Nematodes are also known as

(A) Roundworms (B) Nema

(C) Threadworms (D) **All of the above**

6. The study of free living and plant parasitic nematodes is called as

(A) Plant nematology (B) **Nematology**

(C) Helminthology (D) All of the above

7. The study of nematodes parasitic to man and animals is called as

(A) Plant nematology (B) Nematology

(C) **Helminthology** (D) All of the above

8. The model plant parasitic nematode is

(A) ***Caenorhabditis elegans*** (B) *Meloidogyne* sp.

(C) *Thylenchus* sp. (D) *Pratylenchus* sp.

9. The First plant parasitic nematode whose genome is sequenced is
   (A) ***Caenorhabditis elegans*** (B) *Meloidogyne* sp.
   (C) *Thylenchus* sp. (D) *Pratylenchus* sp.

10. World's most damaging plant parasitic nematode is
   (A) *Heterodera* sp. (B) ***Meloidogyne* sp.**
   (C) *Tylenchus* sp. (D) *Pratylenchus* sp.

11. *Meloidogyne incognita* is a
   (A) Migratory endoparasite (B) Sedentary ectoparasite
   (C) Migratory ectoparasite (D) **Sedentary endoparasite**

12. *Pratylenchus* sp is a
   (A) **Migratory endoparasite** (B) Sedentary ectoparasite
   (C) Migratory ectoparasite (D) Sedentary endoparasite

13. *Tylenchulus* sp is a
   (A) Migratory endoparasite (B) **Sedentary ectoparasite**
   (C) Migratory ectoparasite (D) Sedentary endoparasite

14. *Helicotylenchus* sp is a
   (A) Migratory endoparasite (B) Sedentary ectoparasite
   (C) **Migratory ectoparasite** (D) Sedentary endoparasite

15. Root knot nematode in guava is
   (A) *Heterodera avenae* (B) *Meloidogyne incognita*
   (C) ***Meloidogyne enterlobii*** (D) *Tylechulus semipenetrans*

16. Molya disease of wheat and barley is caused by
   (A) ***Heterodera avenae*** (B) *Meloidogyne incognita*
   (C) *Anguina tritici* (D) *Tylechulus semipenetrans*

17. Ufra disease of Rice is caused by
   (A) *Heterodera avenae* (B) ***Ditylenchus angustatus***
   (C) *Anguina tritici* (D) *Tylechulus semipenetrans*

18. Ufra disease of Rice is more prevalent in
   (A) Transplanted rice (B) Deep water rice
   (C) **Semi-dryrice** (D) Upland rice

19. White tip disease of Rice is caused by
   (A) *Heterodera avenae* (B) *Ditylenchus angustatus*
   (C) *Anguina tritici* (D) ***Aphelenchoides besseyi***

20. Hot water treatment of seedling (50-55°C for 10 min) is recommended for
    (A) *Heterodera avenae* (B) *Ditylenchus angustatus*
    (C) *Anguina tritici* (D) ***Aphelenchoides besseyi***

21. Tobacoo variety reported resistant against RKN is
    (A) ABT1 (B) **ABT10**
    (C) ABT5 (D) None of the above

22. White tip disease of Rice is a
    (A) Soil borne (B) **Seed borne**
    (C) Both A and B (D) None of the above

23. Root knot nematode in Rice, onion and mulberry is
    (A) ***Meloidogyne graminicola*** (B) *Meloidogyne incognita*
    (C) *Meloidogyne arenaria* (D) *Meloidogyne javanica*

24. Seed gall/Ear Cockle disease of wheat is caused by
    (A) *Heterodera avenae* (B) *Meloidogyne incognita*
    (C) ***Anguina tritici*** (D) *Tylechulus semipenetrans*

25. The nematode infesting tea
    (A) ***Pratylenchus loosi*** (B) *Heterodera avenae*
    (C) *Anguina tritici* (D) *Tylechulus semipenetrans*

26. The nematode infesting chickpea
    (A) ***Pratylenchus thornei*** (B) *Heterodera avenae*
    (C) *Anguina tritici* (D) *Tylechulus semipenetrans*

27. Yellow slime/Tundu disease in wheat is caused by the association between
    (A) *Heterodera avenae-+ clavibacter tritici*
    (B) *Anguina tritici + Streptobacillus tritici*
    (C) ***Anguina tritici + Clavibacter tritici***
    (D) *Thylechulus semipenetrans + Clavibacter tritci*

28. The nematode responsible for Slow decline of Dieback disease in citrus is
    (A) *Heterodera avenae* (B) *Meloidogyne incognita*
    (C) *Anguina tritici* (D) ***Tylechulus semipenetrans***

29. The nematode of quarantine importance in India is
    (A) *Heterodera avenae* (B) *Meloidogyne incognita*
    (C) *Anguina tritici* (D) ***Globodera rostochinensis***

30. The first nematode discovered by Needham (1743) was
    (A) *Heterodera avenae* (B) *Meloidogyne incognita*
    (C) ***Anguina tritici*** (D) *Globodera rostochinensis*

31. The Root knot nematode causing gall in green house cucumber was first reported by
    (A) Needham (B) NA Cobb
    (C) Kuhn (D) **Berkley**

32. Father of Nematology is
    (A) Needham (B) **NA Cobb**
    (C) Kuhn (D) Berkley

33. Root knot nematode incidence in Tea in India was first reported by
    (A) Butler (B) Dastur
    (C) **Barber** (D) Siddiqui

34. "Ufra" disease on rice in Bengal caused by *Ditylenchus angustus* was first reported by
    (A) **Butler** (B) Dastur
    (C) Barber (D) Siddiqui

35. White tip of rice caused by *Aphelenchoides besseyi* in central India was first reported by
    (A) Butler (B) **Dastur**
    (C) Barber (D) Siddiqui

36. Citrus nematode for the first time from India was first reported by
    (A) Butler (B) Dastur
    (C) Barber (D) **Siddiqui**

37. Nematicidal value of DD wash was first reported by
    (A) **Carter** (B) FGW Jones
    (C) Barber (D) MRGK Nair

38. Potato cyst nematode for the first time from India was first reported by
    (A) Carter (B) **FGWJones**
    (C) Barber (D) MRGK Nair

39. Burrowing nematode of banana for the first time in India was first reported by
    (A) **Carter** (B) FGW Jones
    (C) Barber (D) **MGRK Nair**

40.Which of the following system is absent in Plant Parasitic Nematodes

(A) Respiratory (B) Circulatory

(C) **Both A and B** (D) Excretory

41. The Plant Parasitic Nematodes are

(A) Triploblastic (B) Un segmented

(C) Bilaterally symmetrical (D) All of the above

42. The shape of the Plant Parasitic Nematodes is

(A) Fusiform (B) Cylinderical

(C) **Vermiform** (D) All of the above

43. Most of the Plant Parasitic Nematodes belong to order

(A) Dorylaimida (B) **Tylenchida**

(C) Aphelenchida (D) Rhabidita

44. Largest Plant Parasitic Nematodes is

(A) ***Placentonema gigantisima*** (B) *Griffiella* sp.

(C) *Meloidogyne* sp. (D) *Paratylenchus* sp.

45. Smallest Plant Parasitic Nematodes is

(A) *Placentonema gigantisima* (B) ***Griffiella* sp.**

(C) *Meloidogyne* sp. (D) *Paratylenchus* sp.

46. Ring disease of Coconut is caused by

(A) *Rhadopholus similis* (B) ***Rhadinaphelenchus cocophilus***

(C) *Meloidogyne* sp. (D) *Paratylenchus* sp.

47. Stylet is absent in

(A) Longidorus (B) **Monochus**

(C) Xiphinema (D) Trichodorus

48. Odontostyle is absent in

(A) Tylenchus (B) Rhabdus

(C) Xiphinema (D) **Dorylaimus**

49. Cardia in plant parasitic nematodes helps in

(A) Passage of food (B) **Absorption**

(C) Regurgitation (D) Digestion

50. Sexual dimorphism well established in plant parasitic nematodes is

(A) *Helicotylenchus* (B) ***Meloidogyne***

(C) *Anguina* (D) *Trichodorus*

51.Perineal pattern is used for the species identification in

(A) *Heterodera* (B) ***Meloidogyne***

(C) *Globodera* (D) *Ditylenchus*

52. Lemon shaped cysts are present in

(A) ***Heterodera*** (B) *Meloidogyne*

(C) *Globodera* (D) *Ditylenchus*

53. All known plant parasitic nematodes possess

(A) **Stylet** (B) Amphid

(C) Spicule (D) Phasmid

54. The number of Juvenile stages in plant parasitic nematodes are

(A) Two (B) Three

(C) **Four** (D) Five

55. The infective Juvenile stages in plant parasitic nematodes is

(A) IJ 1 (B) **IJ 2**

(C) IJ 3 (D) IJ 4

56. The infective Juvenile stages in Entomopathogenic nematodes is

(A) IJ 1 (B) IJ 2

(C) **IJ 3** (D) IJ 4

57. The plant parasitic nematodes reproduce mainly through

(A) Viviparity (B) Polyembrony

(C) **Amphimixis** (D) Parthenogenesis

58. The nematodes pathogenic to insects is

(A) *Rhabditis* (B) *Steinernema*

(C) **Both A and B** (D) None of the above

59. The cuticle of plant parasitic nematodes is secreted by

(A) **Hypodemis** (B) Epidermis

(C) Cortex (D) All of the above

60. The cuticular marking of plant parasitic nematodes is

(A) Ridges (B) Punctuations

(C) Alae (D) **All of the above**

61. All the three Oesophageal glands open in to the medium in

(A) Aphelenchida (B) **Tylenchids**

(C) Rhabtidae (D) *Hetrodera*

62.The Root Knot gelatinous matrix is secreted by
(A) Oesophageal glands (B) Rectum
(C) **Rectalglands** (D) Posteriorbulb

63. Guiding ring is present at the tip of stylet in
(A) Xiphinema (B) **Longidorus**
(C) Tylenchus (D) Dorylaimus

64. The Somatic muscles of plant parasitic nematodes arranged in 2 rows is
(A) Meromyarian (B) **Holomyarian**
(C) Polymyarian (D) All of the above

65.The Somatic muscles of plant parasitic nematodes arranged in 2-5 rows is
(A) **Meromyarian** (B) Holomyarian
(C) Polymyarian (D) All of the above

66. The transmission of plant viruses by nematodes was demonstrated by
(A) Gohenen (B) Raski
(C) Hewitt (D) **All of the above**

67. The mounting media used for preparatoion of permanent mount is
(A) Phenyl (B) DPX
(C) **Glycerine** (D) Lactophenol

68. The vector of NETU virus is
(A) *Xiphinema* (B) *Longidorus*
(C) ***Trichodorus*** (D) *Hoplolaimus*

69. The vector of NEPO virus is
(A) ***Xiphinema*** (B) *Longidorus*
(C) *Trichodorus* (D) *Hoplolaimus*

70. The vector of TOBRA virus is
(A) *Xiphinema* (B) *Longidorus*
(C) ***Trichodorus*** (D) *Hoplolaimus*

71. The Sampling depth for nematode extraction in annual crops is
(A) 10-15cm (B) **15-20cm**
(C) 20-25cm (D) 25-30cm

72. Median bulb in plant parasitic nematodes is a
(A) Secretory (B) Glandular
(C) Globular (D) **Muscular pumping organ**

73. Kanred variety of wheat exhibit resistance against

(A) ***Anguina tritici*** (B) *Heterodera avenae*

(C) *Ditylenchus angustatus* (D) *Pratylenchus loosi*

74. The Threshold population of RKN is approximately

(A) 1 larva /cc soil (B) **2 larva /cc soil**

(C) 3 larva /cc soil (D) 4 larva /cc soil

75. The cysts of *Heterodera avenae* remain viable up to

(A) 1 year (B) **2 year**

(C) 3year (D) 4year

76. The Corpus, Isthmus and Bulb are part of

(A) **Pharynx** (B) Oseophagus

(C) Stylet (D) None of the above

77. Stomatostylet is present in

(A) Dorylaimids (B) **Tylenchids**

(C) Rhabditids (D) Trichodorids

78. The nematode eggs are stored in

(A) Spermatheca (B) Vas deferns

(C) **Post Uterine Sac** (D) Ejaculatory glands

79. The male nematode sperms are stored in

(A) **Spermatheca** (B) Vas deferns

(C) Post Uterine Sac (D) Ejaculatory glands

80. Organ for guiding and directing spicule is called as

(A) Intestine (B) **Gubernaculum**

(C) Cloaca (D) Telamon

81. The oesophageal intestinal valve is called as

(A) Phasmid (B) **Cardia**

(C) Amphid (D) Hemizonids

82. The sensory papillae in cepahlic sensory organs are arranged in order of

(A) **6 inner labial, 6 outer labial and 4 cephalic papillae**

(B) 4 inner labial, 6 outer labial and 4 cephalic papillae

(C) 6 inner labial, 6 outer labial and 6 cephalic papillae

(D) 6 inner labial, 4 outer labial and 4 cephalic papillae

83. The number of sensory papillae in cephalic sensory organs of nematodes is

(A) **16** (B) 32

(C) 08 (D) 48

84. The sensory structure present in anterior and posterior region of cephalic region

(A) **Cephalids** (B) Phasmids

(C) Hemizonids (D) Deirids

85. The pair of small protruberance, one each in the lateral side in the middle of lateral field located in the oesosophageal region

(A) Cephalids (B) Phasmids

(C) Hemizonids (D) **Deirids**

86. Paired sensory structure, one each – lateral side – middle of lateral field opens outside through minute pore

(A) Cephalids (B) **Phasmids**

(C) Hemizonids (D) Deirids

87. Slit like sensory structure present in the lateral side of stoma

(A) Cephalids (B) Phasmids

(C) Hemizonids (D) **Amphids**

88. Environmental conditions due to the lack of water which induce nematodes to enter a quiescent stage is

(A) **Anhydrobiosis** (B) Cryobiosis

(C) Osmobiosis (D) Anoxybiosis

89. Environmental conditions due to lowering temperature which induce nematodes to enter a quiescent stage is

(A) Anhydrobiosis (B) **Cryobiosis**

(C) Osmobiosis (D) Anoxybiosis

90. Environmental conditions due to absence or low levels of oxygen which induce nematodes to enter a quiescent stage is

(A) Anhydrobiosis (B) Cryobiosis

(C) Osmobiosis (D) **Anoxybiosis**

91. Environmental conditions due to absence or low levels of oxygenremoval of water from organisms by a solution that has high osmotic potential which induce nematodes to enter a quiescent stage is

(A) Anhydrobiosis (B) Cryobiosis

(C) **Osmobiosis** (D) Anoxybiosis

92. The term Cyst was first used by
(A) **Chatrin (1887)** (B) Ronald
(C) Doughferty (D) Hewitt

93. intestine of nematode are in origin of
(A) **Endodermal** (B) Ectodermal
(C) Mesodermal (D) None of the above

94. Single chamber/parted oseophagous is present in
(A) **Monochids** (B) Dorlaimids
(C) Tylenchids (D) Mermethids

95. Two chamber/parted oseophagous is present in
(A) Monochids (B) **Dorlaimids**
(C) Tylenchids (D) Mermethids

96. Three chamber/parted oseophagous is present in
(A) Monochids (B) Dorlaimids
(C) **Tylenchids** (D) Mermethids

97. A Single wheat gall may contain *A. tritici* nematodes
(A) **11,000-18,000** (B) 20,000-25,000
(C) 30,000-40,000 (D) None of the above

98. The number of nematodes present in single acre of arable soil
(A) **$3x10^9$** (B) $5x10^9$
(C) $6x10^9$ (D) $10x10^9$

99. The male nematode having single testis is
(A) Diorchic (B) **Monorchic**
(C) Monodelphic (D) Didelphic

100. The female nematode having Two ovaries is
(A) Diorchic (B) Monorchic
(C) Monodelphic (D) **Didelphic**

101. The female nematode having Two ovaries both anteriorly directed is
(A) Didelphic opisthodelphic (B) **Didelphic prodelphic**
(C) Didelphic amphidelphic (D) Monodelphic prodelphic

102. The female nematode having Two ovaries both Posteriorly directed is
(A) **Didelphic opisthodelphic** (B) Didelphic prodelphic
(C) Didelphic amphidelphic (D) Monodelphic prodelphic

103. The female nematode having Two ovaries; one anterior and one posterior
   (A) Didelphic opisthodelphic (B) Didelphic prodelphic
   (C) **Didelphic amphidelphic** (D) Monodelphic prodelphic

104. Example of Migratory Ectoparasitic nematodes is
   (A) *Paratylenchus* (B) *Trichodorus*
   (C) *Criconematoides* (D) **All of the above**

105. Example of Sedentary Ectoparasitic nematodes is
   (A) *Hemicyclophora arenaria* (B) *Cacopaurus pestis*
   (C) **Both A and B** (D) None of the above

106. Example of Migratory Endoparasitic nematodes is
   (A) *Pratylenchus* (B) *Rhadopholus*
   (C) *Hirschmaniella* (D) **All of the above**

107. Example of Sedentary Endoparasitic nematodes is
   (A) *Meloidogyne* (B) *Globodera*
   (C) *Heterodera* (D) **All of the above**

108. Example of Semi Endoparasitic nematodes is
   (A) *Tylenchulus* (B) *Rotylenchus*
   (C) *Naccobus* (D) **All of the above**

109. *Globodera rostichinenis* is under quarantine in which crop
   (A) **Potato** (B) Tomato
   (C) Coconut (D) All of the above

110. *Globodera rostichinenis* is under quarantine at which place
   (A) Kufri (B) **Darjeeling**
   (C) Ooty (D) Hosur

111. Stubby root in corn and sugarcane is caused by
   (A) *Pratylenchus*sp. (B) ***Trichodorus christiei***
   (C) *Belanolaimus longicaudatus* (D) All of the above

112. Root lesion in citrus is caused by
   (A) ***Pratylenchus*sp.** (B) *Trichodorus christiei*
   (C) *Belanolaimus longicaudatus* (D) All of the above

113. Coarse root in corn is caused by
   (A) *Pratylenchus* sp. (B) *Trichodorus christiei*
   (C) ***Belanolaimus longicaudatus*** (D) All of the above

114. Blind disease in strawberry is caused by

(A) *Pratylenchussp.* (B) *Trichodorus christiei*

(C) *Belanolaimus longicaudatus* (D) ***Aphelenchoides frageriae***

115. The citrus nematode is

(A) *Rotylenchus reniformis* (B) ***Tylenchulus semipenetrans***

(C) *Globodera rostochienensis* (D) *Rahdopholus similis*

116. The Banana burrowing nematode is

(A) *Rotylenchus reniformis* (B) *Tylenchulussemipenetrans*

(C) *Globodera rostochienensis* (D) ***Rahdopholus similis***

117. The reniform nematode is

(A) ***Rotylenchus reniformis*** (B) ***Tylenchulus semipenetrans***

(C) *Globodera rostocchienensis* (D) *Rahdopholus similis*

118. The Golden nematode is

(A) *Rotylenchus reniformis* (B) ***Tylenchulus semipenetrans***

(C) ***Globodera rostochienensis*** (D) *Rahdopholus similis*

119. The cyst is formed by

(A) **Female** (B) Cuticle

(C) Rectalglands (D) Pharynx

120. Acyst can survive a maximum of how many number of years

(A) 7 years (B) 20 years

(C) 5 years (D) 50 years

121. Galled seeds can survive a maximum of how many number of years

(A) 27 years (B) **39 years**

(C) 48 years (D) 50 years

122. The Juvenile stage (J1) first moult takes place inside the egg only in

(A) Adenophora (B) **Secernentea**

(C) Mermethids (D) None of the above

123. Outer body tube of nematodes consist of

(A) Cuticle (B) Hypodermis

(C) Muscle layer (D) **All of the above**

124. Cuticular markings which help in locomotion is

(A) **Longitudinal markings** (B) Alae

(C) Caudal Alae (D) Cervical Alae

125. Thickenings or projections of cuticle in lateral or sub lateral region are
(A) Longitudinal markings (B) **Alae**
(C) Caudal Alae (D) Cervical Alae

126. Couplatory bursa-posterior region in male contains
(A) Longitudinal markings (B) Alae
(C) **Caudal Alae** (D) Cervical Alae

127. Anterior portion of cuticular markings is
(A) Longitudinal markings (B) Alae
(C) Caudal Alae (D) **Cervical Alae**

128. The layer below cuticular layer is
(A) Epidermis (B) Cortex
(C) **Hypodermis** (D) Mesodemis

129. Longitudinal thickenings of hypodermis is known as
(A) Longitudinal markings (B) Alae
(C) **Chords** (D) Cervical Alae

130. Muscle cells or fibre are seen in between
(A) Longitudinal markings (B) Alae
(C) **Chords** (D) Cervical Alae

131. The number of lips in stoma of a nematode is
(A) 4 (B) 5
(C) **6** (D) 8

132. The pH of the pseudocoelomic fluid is
(A) Acidic (B) Basic
(C) Alkaline (D) **Neutral**

133. The muscular pharynx or Oseophagous of nematode contain
(A) **4 region** (B) 3 region
(C) 2 region (D) 5 regions

134. The Pore in the spear or stylet is
(A) Hole (B) Invagination
(C) **Lumen** (D) **All of the above**

135. Dorsal glands secreting digestive enzymes open in the median bulb in
(A) Tylenchids (B) **Aphelenchoids**
(C) Monochids (D) All of the above

136. Dorsal glands secreting digestive enzymes open below the base of stylet in

(A) **Tylenchids** (B) Aphelenchoids
(C) Monochids (D) All of the above

137. Rectum in female open through

(A) **Anus** (B) Cloaca
(C) Ileum (D) All of the above

138. Rectum in male open through

(A) Anus (B) **Cloaca**
(C) Ileum (D) All of the above

139. Number of couplatory spicules in nematode is

(A) Only One (B) Two
(C) **One pair** (D) Two pair

140. The basal cuticularize guiding plate found in the genital tract of nematodes is

(A) Bursa (B) Spicules
(C) **Gubernaculum** (D) Spermatheca

141. Vulva is located in

(A) Dorsal region (B) **Ventral region**
(C) Lateral region (D) All of the above

142. Rectal glands and Bursa absent in

(A) Secernentea (B) Phasmidia
(C) **Adenophora** (D) All of the above

143. H or U shaped excretory system found in

(A) **Secernentea** (B) Phasmidia
(C) Adenophora (D) All of the above

144. Excretory system found in Tylenchida is

(A) **Canalicular** (B) Glandular
(C) Vesicular (D) H-shaped

145. Male tail curved in

(A) *Longidorus* (B) ***Trichodorus***
(C) *Paratrichodorus* (D) *Xiphinema*

146. Enlarged specialised plant cells formed by the dissolution of the cell wall

(A) **Syncytia** (B) Giantcell
(C) Multinucleate transfer cells (D) All of the above

147. Enlarged specialised plant cells formed by cell division but without compartmentalization of cells is known as

(A) Syncytia (B) **Giant cell**

(C) Multinucleate transfer cells (D) All of the above

148. Syncytium are formed in plants as a result of feeding by

(A) ***Heterodera* sp.** (B) *Meloidogyne* sp.

(C) *Anguina* sp. (D) *Diytlenchus* sp.

149. Giant cells are formed in plants as a result of feeding by

(A) *Heterodera* sp. (B) ***Meloidogyne* sp.**

(C) *Anguina* sp. (D) *Diytlenchus* sp.

150. Fan leaf virus of grape is transmitted by

(A) ***Xiphinema index*** (B) *Xiphinema americanum*

(C) *Triuchodorus pachydermus* (D) All of the above

151. Bud mosaic virus of Peach is transmitted by

(A) *Xiphinema index* (B) ***Xiphinema americanum***

(C) *Triuchodorus pachydermus* (D) All of the above

152. Rattle leaf virus of grape is transmitted by

(A) *Xiphinema index* (B) *Xiphinema americanum*

(C) ***Triuchodorus pachydermus*** (D) All of the above

153. Fan leaf virus of grape is transmitted by

(A) ***Xiphinema index*** (B) *Xiphinema americanum*

(C) *Triuchodorus pachydermus* (D) All of the above

154. Root knot nematode found in chillies, tomato, groundnut

(A) *Meloidogyne incognita* (B) *Meloidogyne hapla*

(C) *Meloidogyne javanica* (D) ***Meloidogyne arenaria***

155. Root knot nematode found in cotton

(A) ***Meloidogyne incognita*** (B) *Meloidogyne hapla*

(C) *Meloidogyne javanica* (D) *Meloidogyne arenaria*

156. Root knot nematode found in sugarcane

(A) *Meloidogyne incognita* (B) *Meloidogyne hapla*

(C) ***Meloidogyne javanica*** (D) *Meloidogyne arenaria*

157. Sugar beet cyst nematode is

(A) ***Heterodera schachti*** (B) *Heterodera trifolii*

(C) *Heterodera gothinjiana* (D) *Heterodera cruciferae*

158. Carrot cyst nematode is
(A) *Heterodera schachti* (B) *Heterodera trifolii*
(C) *Heterodera gothinjiana* (D) ***Heterodera carotae***

159. Mentek disease of rice is caused by
(A) *Heterodera schachti* (B) *Heterodera trifolii*
(C) *Heterodera gothinjiana* (D) ***Hirschmanniella oryzae***

160. The maximum number of larvae found in the wheat gall is
(A) 3600 (B) 5000
(C) 10,000 (D) **32400**

161. Pepper yellow disease is caused by
(A) *Heterodera schachti* (B) *Heterodera trifolii*
(C) *Ditylenvchus* sp. (D) ***Rhadopholus similis***

162. Black Toppling disease of banana is caused by
(A) *Heterodera schachti* (B) *Heterodera trifolii*
(C) *Ditylenvchus* sp. (D) ***Rhadopholus similis***

163. Red ring disease of coconut is caused by
(A) *Heterodera schachti* (B) *Heterodera trifolii*
(C) ***Rhadinaphelechus cocophilus*** (D) *Rhadopholus similis*

164. Nematode wool is a surviving mechanism found in
(A) *Heterodera schachti* (B) *Heterodera trifolii*
(C) ***Ditylenvchus dipsaci*** (D) *Rhadopholus similis*

165. Stunt nematode is
(A) ***Tylenchorhynchus*** (B) *Hoplolaimus*
(C) *Helicotylenchus* (D) *Xiphinema*

166. Lance nematode is
(A) *Tylenchorhynchus* (B) ***Hoplolaimus***
(C) *Helicotylenchus* (D) *Xiphinema*

167. Spiral nematode is
(A) *Tylenchorhynchus* (B) *Hoplolaimus*
(C) ***Helicotylenchus*** (D) *Xiphinema*

168. Dagger nematode is
(A) *Tylenchorhynchis* (B) *Hoplolaimus*
(C) *Helicotylenchus* (D) ***Xiphinema***

169.Leaf gall of strawberry is caused by

(A) *Heterodera schachti* (B) *Heterodera trifolii*

(C) ***Ditylenvchus dipsaci*** (D) *Rhadopholus similis*

170. Decanting and sieving method for nematode extraction was given by

(A) **NA Cobb** (B) Sheshadri

(C) Butler (D) Pradhan

171. Tuberose cultivar Prajwal is resistant to

(A) *R. similis* (B) *M. incognita*

(C) *H. avenae* (D) ***A. besseyi***

172. Nematicidal properties of Ethylene Bromide was first reported by

(A) **Christie** (B) Carter

(C) Kuhn (D) Butler

173. Which of the following is a true Nematicide

(A) Nimitz ZT M (B) Manna

(C) Fluensulfone (D) **All of the above**

174. Paring and hot water treatment of suckers (55°C for 20min) is recommended for the control of

(A) ***R. similis*** (B) *M. incognita*

(C) *H. avenae* (D) *A. besseyi*

175. Chillies cultivar resistant to *M. incognita*

(A) NP 46A (B) Pusa Jawala

(C) **Both A and B** (D) None of the above

176. Tomato cultivar resistant to *M. incognita*

(A) PNR-7 (B) Hisar Lalit

(C) NDTR D) **All of the above**

177. Potato cultivar resistant to *Globodera rostochinensis* is

(A) Kufri Pukhraj (B) Kufri Chipsona

(C) **Kufri Swarna** (D) Kufri Chandermukhi

178. Barley cultivar resistant to *Heterodera aveanae* is

(A) BH-75 (B) C-164

(C) Rajkiran (D) **All of the Above**

179. Grapes cultivar resistant to *Globodera rostochinensis* is

(A) Dodrige (B) 1613

(C) Salt Creek (D) **All of the above**

180. Nematode egg parasitic fungi is

(A) ***Papecilomyces lilacinus*** (B) *Pochonia chlaymdosporia*

(C) *Pseudomonas florescens* (D) *Trichoderma sp.*

181. The minimum limit of ear cockle disease in wheat for foundation and certified seed category of seed production is

(A) **Zero per cent** (B) One per cent

(C) 1-2 per cent (D) 2 per cent

182. Odontostylet is present in

(A) **Dorylaimida** (B) Triplonchida

(C) Tylenchida (D) Aphelenchida

183. Onchiostylet is present in

(A) Dorylaimida (B) **Triplonchida**

(C) Tylenchida (D) Aphelenchida

184. Stomatostylet is present in

(A) Tylenchida (B) Aphelenchida

(C) **Both A and B** (D) Triplonchida

185. The Nematodes pathogenic to insects is

(A) *Heterorhabditis indica* (B) *Steinernema carpocapsae*

(C) *Heterorhabditis bacteriophora* (D) **All of the above**

186. The DD-36 is a strain of

(A) *Heterorhabditis indica* (B) ***Steinernema carpocapsae***

(C) *Heterorhabditis bacteriophora* (D) All of the above

187. The entomopathogenic nematode used a biocontrol agent against mosquitoes is

(A) *Heterorhabditis indica* (B) *Steinernema carpocapsae*

(C) *Heterorhabditis bacteriophora* (D) ***Romanomermis culicivorax***

188. The bacterium associated with *Steinernema carpocapsae* is

(A) ***Xenorhabdus* sp.** (B) *Photorhabdus* sp.

(C) Both A and B (D) none of the above

189. The bacterium associated with *Heterorhabditis indica* is

(A) *Xenorhabdus* sp. (B) ***Photorhabdus* sp.**

(C) Both A and B (D) None of the above

190. The infected host turns brick red and luminesces in the dark is the symptom of attack of which nematode

(A) ***Heterorhabditis indica*** (B) *Steinernema carpocapsae*

(C) *Romanomermis culicivorax* (D) All of the above

191. *The infective stage of entomopathogenic nematodes is*

(A) J1 (B) J2

(C) **J3** (D) J4

192. The rate of application of *Steinernemacarpocapsae* for cutworm control is

(A) 2 billion IJs/ha (B) 3 billion IJs/ha

(C) 4 billion IJs/ha (D) **5 billion IJs/ha**

193. The entomopathogenic nematodes kill the hosts in

(A) **24-48 hours** (B) 48-72 hours

(C) 72-96 hours (D) >92 hours

194. The Nemagel is a commercial formulation of

(A) *Heterorhabditis indica* (B) *Steinernema carpocapsae*

(C) *Heterorhabditis bacteriophora* (D) ***Steinernema thermophilum***

195. The NemaStar is a commercial formulation of

(A) *Heterorhabditis indica* (B) ***Steinernema carpocapsae***

(C) *Heterorhabditis bacteriophora* (D) *Steinernema thermophilum*

196. The Ecomask is a commercial formulation of

(A) *Heterorhabditis indica* (B) ***Steinernema carpocapsae***

(C) *Heterorhabditis bacteriophora* (D) *Steinernema thermophilum*

197. The BioVector is a commercial formulation of

(A) *Heterorhabditis indica* (B) ***Steinernema carpocapsae***

(C) *Heterorhabditis bacteriophora* (D) *Steinernema thermophilum*

198. The Nemasys is a commercial formulation of

(A) *Heterorhabditis indica* (B) *Steinernema carpocapsae*

(C) *Heterorhabditis bacteriophora* (D) ***Steinernema feltie***

199. The NemaPlus is a commercial formulation of

(A) *Heterorhabditis indica* (B) *Steinernema carpocapsae*

(C) *Heterorhabditis bacteriophora* (D) ***Steinernema feltie***

200. The Etonem is a commercial formulation of

(A) *Heterorhabditis indica* (B) *Steinernema carpocapsae*

(C) *Heterorhabditis bacteriophora* (D) ***Steinernema feltie***

201. The Proactant is a commercial formulation of
   (A) ***Steinernema scapterisci*** (B) *Steinernema carpocapsae*
   (C) *Heterorhabditis bacteriophora* (D) *Steinernema feltie*

202. The GrubStake is a commercial formulation of
   (A) *Steinernema scapterisci* (B) *Steinernema carpocapsae*
   (C) ***Heterorhabditis bacteriophora*** (D) *Steinernema feltie*

203. The Nema Green is a commercial formulation of
   (A) *Steinernema scapterisci* (B) *Steinernema carpocapsae*
   (C) ***Heterorhabditis bacteriophora*** (D) *Steinernema feltie*

204. The LawnPatrol is a commercial formulation of
   (A) *Steinernemascapterisci* (B) *Steinernema carpocapsae*
   (C) ***Heterorhabditis bacteriophora*** (D) *Steinernema feltie*

205. The Soil Commandoes is a commercial formulation of
   (A) *Steinernema scapterisci* (B) *Steinernema carpocapsae*
   (C) ***Heterorhabditis bacteriophora*** (D) *Steinernema feltie*

206. The entomopathogenic nematode enter host through integument
   (A) ***Heterorhabditis indica*** (B) *Steinernema carpocapsae*
   (C) *Steinernema thermophilum* (D) All of the above

207. The entomopathogenic nematode enter host through integument and natural opening of the host
   (A) ***Heterorhabditis indica*** (B) *Steinernema carpocapsae*
   (C) Both A and B (D) None of the above